T0192512

What Every Engineer Should Know About Smart Cities

Get ready to be at the forefront of the future of urban development!

As cities continue to rapidly grow, the demand for sustainable and efficient infrastructure becomes more urgent. That's where *What Every Engineer Should Know About Smart Cities* comes in, offering a comprehensive guide to the concepts and technologies driving the transformation of our cities.

Delve into the world of smart cities and discover how information and communication technologies are revolutionizing urban environments. With clear definitions and a focus on real-world applications, this book explores the benefits and challenges of smart cities. It also highlights interdisciplinary topics such as smart buildings, autonomous cars, and urban emergency management systems.

This book is not just a theoretical exploration of smart cities. It goes beyond that by providing an in-depth look at the key technologies that are essential to creating smart cities. From the Internet of Things and blockchain to digital twins and modeling and simulations, readers will gain a solid understanding of the foundational technologies that make smart cities possible. With detailed discussions and real-world examples of smart mobility, smart health, smart education, and smart agribusiness, readers will gain a deep understanding of the requirements and characteristics that engineers need to contribute to the development of smart cities.

Whether you're an engineer looking to expand your knowledge, a city planner seeking to understand the latest trends, or simply someone interested in the future of urban living, *What Every Engineer Should Know About Smart Cities* is the ultimate guide to unlocking the potential of smart cities for sustainable urban development and improved quality of life.

What Every Engineer Should Know
Series Editor: Phillip A. Laplante, Pennsylvania State University

Series Statement

What every engineer should know includes an overwhelming catalogue of information. Regardless of discipline, engineering intersects most scientific fields and modern technologies. Practicing engineers however must also navigate managerial, socio-economic, and even political concerns. The engineer discovers soon after graduation that any curriculum omits important and thorny issues of daily practice—for example, problems concerning new technologies, scientific advances, business practices, legal implications, and team dynamics.

With the *What Every Engineer Should Know* series of concise, easy-to-understand volumes, every engineer can access primers on important subjects across a broad range of knowledge areas including intellectual property, contracts, software, business communication, management science, and risk analysis, as well as more specific topics such as embedded systems design. These books are very accessible to every engineer, scientist, and technology professional and are necessary to remain competitive in this dynamic, global economy.

What Every Engineer Should Know about Modeling and Simulation
Raymond J. Madachy and Daniel Houston

What Every Engineer Should Know about Excel, Second Edition
J.P. Holman and Blake K. Holman

Technical Writing: A Practical Guide for Engineers, Scientists, and Nontechnical Professionals, Second Edition
Phillip A. Laplante

What Every Engineer Should Know About the Internet of Things
Joanna F. DeFranco and Mohamad Kassab

What Every Engineer Should Know about Software Engineering
Phillip A. Laplante and Mohamad Kassab

What Every Engineer Should Know About Cyber Security and Digital Forensics
Joanna F. DeFranco and Bob Maley

Ethical Engineering: A Practical Guide with Case Studies
Eugene Schlossberger

What Every Engineer Should Know About Data-Driven Analytics
Phillip A. Laplante and Satish Mahadevan Srinivasan

What Every Engineer Should Know About Reliability and Risk Analysis
Mohammad Modarres and Katrina Groth

What Every Engineer Should Know About Smart Cities
Valdemar Vicente Graciano Neto and Mohamad Kassab

For more information about this series, please visit: www.routledge.com/What-Every-Engineer-Should-Know/book-series/CRCWEESK

What Every Engineer Should Know About Smart Cities

Valdemar Vicente Graciano Neto
and Mohamad Kassab

CRC Press
Taylor & Francis Group
Boca Raton London New York

CRC Press is an imprint of the
Taylor & Francis Group, an **informa** business

Designed cover image: © Shutterstock

MATLAB® is a trademark of The MathWorks, Inc. and is used with permission. The MathWorks does not warrant the accuracy of the text or exercises in this book. This book's use or discussion of MATLAB® software or related products does not constitute endorsement or sponsorship by The MathWorks of a particular pedagogical approach or particular use of the MATLAB® software.

First edition published 2024
by CRC Press
6000 Broken Sound Parkway NW, Suite 300, Boca Raton, FL 33487-2742

and by CRC Press
4 Park Square, Milton Park, Abingdon, Oxon, OX14 4RN

CRC Press is an imprint of Taylor & Francis Group, LLC

© 2024 Valdemar Vicente Graciano Neto and Mohamad Kassab

ISBN: 978-1-032-39093-2 (hbk)
ISBN: 978-1-032-39136-6 (pbk)
ISBN: 978-1-003-34854-2 (ebk)

DOI: 10.1201/9781003348542

Typeset in Times
by KnowledgeWorks Global Ltd.

Contents

Preface

In the 1980s and 1990s, we marveled at The Jetsons, a TV show portraying a typical family living with robots that managed their daily tasks, and flying cars that transported them. The family used telemedicine and various electronic equipment to automate their lives, and what stood out was their extensive use of technology to enhance their household and urban environments. Today, this is becoming a reality through the rise of "smart cities."

Although futuristic predictions are frequently exaggerated, current smart city projects are making considerable progress in achieving the functionalities and automation seen in The Jetsons. While some technologies, such as telemedicine and cleaning robots, are already accessible, prototypes for electric and flying cars are in development. Building smart cities necessitates a wide range of skilled professionals who can design and execute the necessary civil architecture, transportation strategies, and smarter citizens who can effectively use available resources.

This book provides engineers with an extensive understanding of smart cities and the technologies involved in making urban environments smarter. It delivers an immersive experience and insight into the requirements and features that engineers need to know to contribute to the engineering of smart cities. This book also examines how emerging technologies such as the Internet of Things, blockchain, and artificial intelligence are linked to the concepts and subdomains of smart cities. The language used is easy to understand so that any engineer can grasp the concepts and apply them in their professional life.

Whether you are a civil, computer, software, systems, electrical, mechanical, automobile, automation, production, or traffic engineer, or a student, academician, or practitioner, this book is a quick guide to acquiring foundational knowledge of smart city concepts. It provides a comprehensive overview of the technologies involved in building smart cities.

To make the reading experience more engaging, each chapter includes supplementary discussions on important topics in gray rectangles. These discussions cover a range of topics, including pop culture references, TV shows, news, specific discussions for different branches of engineering, and recommended literature that offers unique perspectives on smart cities. We hope this book will equip you with the essential knowledge and tools to navigate the ever-evolving landscape of smart cities, and we wish you an enjoyable reading experience.

Notes on Referencing and Errors

The authors have tried to uphold the highest standards for giving credit where credit is due. Each chapter contains a list of related readings, which have been extensively updated for the fourth edition, and they should be considered the primary references for that chapter. Where direct quotes or non-obvious facts are used, an appropriate note or in-line citation is provided. In particular, it is noted where the authors published portions in preliminary form in other scholarly publications.

Despite these best efforts and those of the reviewers and publisher, there are still likely errors to be found. Therefore, if you believe that you have found an error – whether it is a referencing issue, factual error, or typographical error, please contact the authors at: valdemarneto@ufg.br (Valdemar Vicente Graciano Neto) or muk36@psu.edu (Mohamad Kassab).

Disclaimer

When discussing certain proprietary systems, every effort has been taken to disguise the identities of any organizations as well as individuals who are involved. In these cases, the names and even elements of the situation have been changed to protect the innocent (and guilty). Therefore, any similarity between individuals or companies mentioned herein is purely coincidental.

Acknowledgments

We would like to express our gratitude to the many individuals who have assisted us during the preparation of this book. In particular, we wish to extend our thanks to Elisa Yumi Nakagawa for her significant contributions to Chapter 3 on systems of systems and Chapter 5 on digital twins, as well as to Rodrigo Pereira dos Santos and Wallace Manzano for their contributions to Chapter 3 on systems of systems and Leonardo Vieira Barcelos to his contributions to Chapter 5 on digital twins. We also want to thank Lina Garcés for the great contribution to Chapter 8 on building healthy cities and Jonas Gomes, José Maria David, Wagner Arbex, and Regina Braga for their essential contributions to Chapter 9 on agribusiness. Additionally, we are grateful to Rafael Zancan Frantz, Sandro Sawicki, Gerson Battisti, Fabricia Roos Frantz, Pedro Henrique Dias Valle, and Odaylson Elder for their major contributions to the case study in Chapter 11.

We would also like to take a moment to express our personal gratitude.

Dr. Graciano Neto would like to acknowledge Prof. Dr. Antonio Carlos de Oliveira Junior and his group, as well as Romulo França and Dr. Renato Bulcao-Neto, for their help and support. He is thankful to his family and friends for their encouragement and support, particularly his lovely mother (in memoriam).

Dr. Kassab would like to extend his appreciation to his family members Hassan, Mariam, Essa, Eman, Sana, and Samar for their support and love during the journey of writing this book.

We are indebted to so many and are grateful for their contributions and support.

About the Authors

Dr. Valdemar Vicente Graciano Neto is Associate Professor with tenure at the Informatics Institute (INF) of the Federal University of Goiás (UFG) in Goiânia, Brazil, since 2014. He is currently the Coordinator of the Information Systems bachelor's degree course at INF-UFG from 2021 to 2023. Dr. Valdemar holds a Ph.D. double degree in Computer Science and Computational Mathematics from the Institute of Mathematics and Computing Sciences of the University of São Paulo (ICMC-USP) and in Science and Information Technology with a focus on Informatics from the University of South Brittany, France. With more than 10 years of teaching experience, he is a member of the Graduate Program in Computer Science at UFG, supervising both master's and doctoral projects.

He served as the Program Committee Chair for SBSI 2022 and Coordinator of the Steering Committee for the Brazilian Symposium on Information Systems 2023. In addition, he was a member and coordinator of the Special Committee on Information Systems (CE-SI) of the Brazilian Computing Society (SBC) from 2015 to 2017, 2018 to 2019, and 2021 to 2024. Dr. Valdemar has expertise in Software Engineering and Information Systems, with a focus on Software Development and Architecture. He has specific knowledge in Model-Based Software Engineering, Systems-of-Systems, Smart Cities, and Information Systems Engineering. He was also a representative member of the Computer Science area in the Area Advisory Committee (CAA) for the National Student Performance Examination (ENADE 2020) at the invitation of INEP, a division of the Brazilian Ministry of Education.

He has published more than 100 papers and articles in conferences and journals, as well as book chapters. Dr. Valdemar has contributed to writing chapters for the *Body of Knowledge of Modeling and Simulation*, an initiative of The Society for Modeling and Simulation International. He is also a member of the SBC.

Dr. Mohamad Kassab is an associate professor and a member of the graduate faculty at The Pennsylvania State University. He earned his Ph.D. and M.S. degrees in Computer Science from Concordia University in Montreal, Canada. Dr. Kassab was an affiliate assistant professor in the Department of Computer Science and Software Engineering at Concordia University between 2010 and 2012 and a postdoctoral researcher in Software Engineering at Ecole de Technologie Supérieure (ETS) in Montreal between 2011 and 2012, and a visiting scholar at Carnegie Mellon University (CMU) between 2014 and 2015.

Dr. Kassab has been conducting research projects jointly with the industry to develop formal and quantitative models to support the integration of quality requirements within software and systems development life cycles. The models are being further leveraged with the support of developed architectural frameworks and tools. His research interests also include bridging the gap between software engineering practices and disruptive technologies (e.g., IoT, blockchain). He has published extensively in software engineering books, journals, and conference proceedings. He is also a member of numerous professional societies and program committees, and the organizer of many software engineering workshops and conference sessions. He authored many books including two other books in this series: *What Every Engineer Should Know About the Internet of Things* and *What Every Engineer Should Know About Software Engineering.*

With more than 20 years of global industry experience, Dr. Kassab has developed a broad spectrum of skills and responsibilities in many software engineering areas. Notable experiences includes business unit manager at Soramitsu, senior quality engineer at SAP, senior quality engineer at McKesson, senior associate at Morgan Stanley, senior quality assurance specialist at NOKIA, and senior software developer at Positron Safety Systems. He is an Oracle Certified Application Developer, Sun Certified Java Programmer, and Microsoft Certified Professional.

Dr. Kassab has taught a variety of graduate and undergraduate software engineering and computer science courses at Penn State and Concordia University. He has won many awards for his excellence in teaching.

1 Defining Smart Cities
An Overview

1.1 INTRODUCTION

The world population is continuously growing. Figure 1.1 illustrates the estimation of evolution of urban population between 2022 and 2050. The World Development Bank stated that, by 2022, 56% of the 8 billion people who lived in the world – 4.4 billion inhabitants – were living in cities. United Nations (UN), in turn, predicted through the World Population Prospects Report in 2022 that the world's population could grow to around 8.5 billion by 2030 and reach 9.7 billion people by 2050 (United Nations, 2022), when almost 70% of the world population would live in urban areas (United Nations, 2019). Reports also estimated that, by 2030, the world can reach 43 so-called megacities (up from 31 today, according to reports) – those with more than 10 million inhabitants – with most of them in developing countries. Given these impacting numbers, the accommodation of the population in urban centers is evidently a complex problem to be solved.

This significant increase in urban population have brought with it a host of challenges, including housing, traffic, food safety, and waste treatment. According to the UN, cities currently account for over 70% of global energy consumption and are responsible for more than 70% of carbon emissions. Furthermore, by 2050, it is expected that urban areas will generate about 2.5 billion tons of solid waste per year, or about 68% of the world's total waste. As such, sustainable urban development is crucial in addressing these challenges and reducing the environmental impact of urbanization.

Consequently, the already crowded and chaotic urban environments should be further improved and prepared to accommodate, in a sustainable, feasible, and balanced way, the higher number of forthcoming citizens. The Kyoto Protocol was established in 1997 and started to be operationalized in 2005. The Kyoto Protocol requires a commitment of industrialized countries and economies to reduce greenhouse gases (GHG) emissions, aka *carbon units*, according to agreed individual targets. Since urban environments are the source of a significant part of the GHG emissions on Earth, a full conformance with Kyoto Protocol requires a city to be more sustainable, that is, the urban environment should be prepared to enable citizens to live their daily routine (going to work, feeding, exercising, working, and transiting) while still preserving the surrounding environment.

Achieving sustainability in cities requires sophisticated solutions, as the interdependence among various aspects of urban life can be highly complex, necessitating smarter approaches. To achieve this, urban objects must be embedded with software

DOI: 10.1201/9781003348542-1

World Population

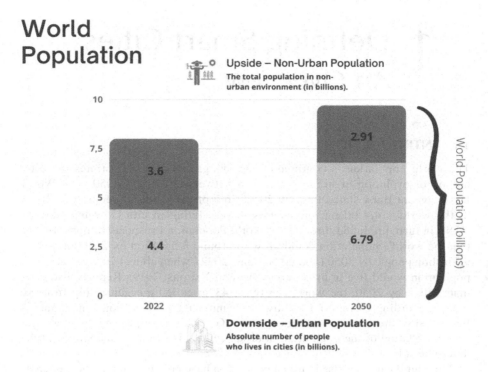

Upside – Non-Urban Population
The total population in non-urban environment (in billions).

World Population (billions)

Downside – Urban Population
Absolute number of people who lives in cities (in billions).

FIGURE 1.1 Estimation of evolution in world population.

and hardware technologies, facilitating communication and enhancing quality of life, including the monitoring and management of carbon emissions from public transportation, optimization of energy consumption, waste management, and other crucial facets of urban living.

The utilization of technology to enhance not only the sustainability objectives but also the overall quality of life in urban areas has prompted the urban and industrial development endeavors to create a new generation of cities known as *smart cities*. These cities are characterized by the integration and interconnectivity of their physical, social, and business infrastructure using Information and Communication Technology (ICT) to optimize the collective services to the community (Harrison et al., 2010; Mohanty et al., 2016).

Over the years, technology has been increasingly integrated into mechanical, electrical, hydraulic, and pneumatic systems, with software supporting them to offer higher precision and automation of operation, making them smarter (Graciano Neto, 2018). Examples of this integration can be seen in everyday devices such as smartphones and smart watches. Similarly, other aspects of urban life have also become "smarter," meaning that they have heavily relied on technology infrastructure, particularly software (Rocha et al., 2019).

The incorporation of technology into smart cities has the potential to offer several advantages for the populace, including improved and cost-effective transportation alternatives, as demonstrated in Singapore (Asiag, 2020); collision avoidance between vehicles, buildings, and pedestrians through inter-system communication to ensure collective safety, as proposed by Taylor et al. (2020); sharing of data and energy to prevent waste, such as sensors measuring energy consumption in a neighborhood and alerting users to conserve power; and employment of artificial intelligence (AI) algorithms to provide more optimized outcomes than ad-hoc strategies, such as more intelligent routing for transportation; just to give a few examples.

Thus, the implementation of smart cities necessitates extensive use of technology in all public sectors. The impact of technology on our daily lives is already evident through mobile applications for transportation and urban private transport. The potential for technology to revolutionize urban life while preserving the environment and expanding urbanization is vast, and scientific advancements continue to offer solutions in various aspects.

This book provides an overview of smart cities, including definitions, key components, and technologies, as well as examples of smart city initiatives in different domains and locations worldwide. In-depth discussions of the challenges facing smart cities and current research efforts in this field will also be presented.

1.2 IMPORTANT DEFINITIONS

The term "smart city" originated in the late 1990s (Yin et al., 2015). It was first used by IBM in its "smart cities" initiative, which aimed to use technology to improve the efficiency and sustainability of urban areas. The idea behind the concept is to use advanced technology and data analysis to address the challenges of urbanization, such as traffic congestion, pollution, and access to services. Other companies and organizations also began using the term to describe their own initiatives, and it has since become a popular buzzword used to describe the integration of technology and data in urban environments.

Giffinger et al. (2007) defined smart city as having six main dimensions: smart economy, smart people, smart governance, smart mobility, smart environment, and smart living.

For the purpose of this book, we understand smart cities as cities having a set of several and non-exhaustive list of smart dimensions. Smart cities are composed of a network of interconnected systems, including smart infrastructure, smart living (including smart buildings, houses, and neighborhoods), smart transportation, smart energy, smart healthcare, smart governance, smart education, and smart citizens, as shown in Figure 1.2. These systems must be able to interoperate at a certain level in order to function effectively. For example, smart traffic control systems and autonomous cars need to be able to communicate with each other, while smart buildings may not have a direct need to interact with traffic control systems.

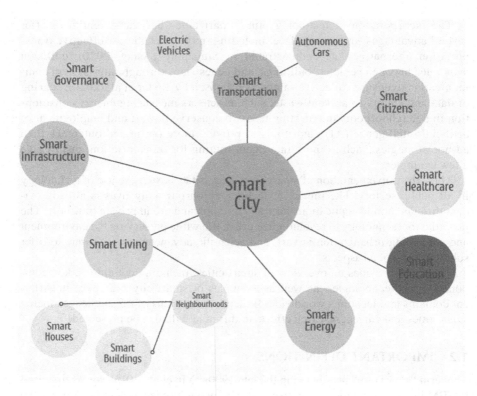

FIGURE 1.2   Smart city dimensions.

TWO CROSSCUTTING CONCERNS: ARCHITECTURE AND INTEROPERABILITY

As you read this book, you will repeatedly see two terms being used under different contexts: architecture and interoperability. These terms crosscut almost the entire book, since interoperability is a key concept for smart cities and architecture is used in a diversity of contexts: a building architecture, a city architecture, a network architecture, a software architecture, a systems-of-systems architecture, and so on.

Architecture is an abstraction whose initial intuition comes from the concept of architecture inherited from the field of "civil construction." In that domain, architecture concerns visual forms and what it expresses. From that perspective, the architecture helps one to distinguish the origin of a given construction. For instance, Greco-Roman architecture is usually made up of large pillars and temples with triangle-shaped facades. On the other hand, Arab architecture already has more rounded shapes, with vaults. Gothic architecture, in turn, has imposing buildings, with tall, pointed towers that try to reach the sky and show the smallness of man in the face of the immensity of what is divine.

The first reference to "software architecture" occurred in 1969 at a conference on software engineering techniques organized by NATO (Kruchten et al., 2006). The idea of aligning architecture and software was recalled in seminal papers in the 1990s (Perry and Wolf, 1992). Before that, the concepts of modularization and separation of interests were already advocated by Dijkstra in the 1970s (Shaw and Clements, 2006) and by the architect and urbanist Christopher Alexander in the late 1970s, who wrote a book entitled "A Pattern Language" (Alexander, 1977) in which he showed how to describe an architecture using patterns.

In summary, herein, the term "architecture" refers to "the structure of something (the main parts and how they are linked, forming a type of topology)" and potentially the behavior that can be obtained from that set of elements. Then, a software architecture corresponds to the main parts of a software, how they are linked, and the behaviors that emerge from the communication between the parts. A hardware architecture is the set of components that form the hardware and enables it to work. A network architecture corresponds to the elements (routers, gateways, and links) that make up that network and the communication that can be obtained from it. The same idea can be analogously replicated to describe a smart city architecture, a system architecture, and so on.

In turn, ISO/IEC defines interoperability as "the ability that two or more systems have to exchange data and use the data exchanged" (ISO/IEC, 2017). This is evidently very important for smart cities, since some functionalities are only possible as the result of the communication between the involved systems. For instance, the behavior "collision avoidance" can be obtained from sensors in an autonomous car, but if the close cars communicate between themselves, the precision of this behavior can be reinforced. Moreover, the communication between sensors and a gateway can alert population of an imminent flood.

Then, over this book, every time you see the terms architecture or interoperability, you can come back here and read it again, if needed.

Standards for smart cities and their components, such as smart grids, Internet of Things (IoT), eHealth, and intelligent transportation systems (ITS), have been developed. One example is ISO 37120 (ISO, 2014), which defines 100 city performance indicators, including economy, education, energy, and environment. IoT plays a crucial role in collecting data for these indicators, such as air quality and energy expenditure, in a neighborhood (IoT is discussed in detail in Chapter 2).

A smart city is made up of various levels of interconnected systems, as depicted in Figure 1.3. These levels are generally defined by geographic coverage. At the most basic level, we have isolated systems such as smart houses, smart buildings, autonomous cars, and smart hospitals. These systems are mainly private properties.

The next level is the communication and connection between these isolated systems, also known as interoperability. For example, a smart home may be linked to a

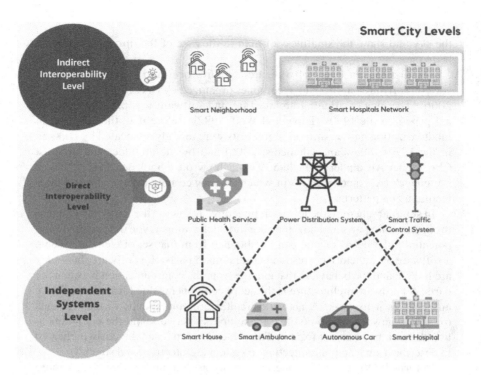

FIGURE 1.3 Illustration of smart city levels.

public emergency service. Additionally, multiple systems within a smart home environment can be linked to different external systems, such as wearable devices linked to emergency services and electrical devices linked to a home power distribution management system, which is part of a public or private smart grid. Each of these smart environments can be considered independent systems that interoperate with other external and independent systems.

As we move up to the next level, we see the formation of a smart city architecture through the indirect interoperability of several systems. For instance, the entire network formed by the smart homes connected to the smart hospitals via the emergency service. This is how the overall shape of the city is formed through the multiple interoperability links established among its constituent systems.

Finally, smart cities should be *sustainable*, a term used more than once herein and in other materials. The concept of sustainability is tightly (and recurrently) related to smart cities. The following gray box brings a brief discussion on the term.

SMART CITIES AND SUSTAINABILITY

Sustainability can be defined under multiple perspectives. Straightforwardly speaking, it is related to the environment preservation. Then, it refers to advance the world, industry, and living with the minimum impact on the environment.

Dictionaries appeal for sustainability as the avoidance of the depletion of natural resources in order to maintain an ecological balance. In summary, it consists of fulfilling the needs of current generations without compromising the needs of future generations, while ensuring a balance between economic growth, environmental care, and social well-being. Then, the development of smart cities comprises the establishment of practices to make the urban living more sustainable, avoiding the impacts that human living inherently has on the environment, and even improving the surrounding environment. Other definitions of sustainability also prescribe the meaning of sustaining at large, including a diversity of dimensions, such as economic viability, environmental protection, and social equity. Hence, sustainability is about preserving the world in multiple dimensions to promote equality and dignity.

1.3 SYSTEMS OF SMART CITIES

A smart city comprises a variety of interconnected and interdependent systems, each of which is "smart" in its own right. These include following.

1.3.1 SMART BUILDINGS

Smart buildings utilize technology to enhance energy efficiency, safety, and security. They gather data from sensors and IoT systems to improve the experience of residents and visitors, offering features such as automated temperature and light control. These buildings often consist of sub-systems such as a fire system for controlling fire sprinklers and issuing alarms in building areas, a lighting system for controlling the light intensity in building areas, and a room system comprising private spaces with sensors for smoke, temperature, and presence. Additional systems such as security cameras and temperature control systems may also be integrated into the smart building system. Studies by Manzano et al. (2020) and Graciano Neto et al. (2019) provide further insight into smart building technology.

1.3.2 SMART HOUSES/HOMES

A smart home utilizes advanced technology to enhance the comfort, convenience, and security of its residents. Key components of a smart home include:

- **Home automation system:** A central control system that connects and controls all the smart devices in the home, allowing residents to manage lighting, temperature, security, and other aspects with a single application or voice commands.
- **Smart thermostat:** A device that enables remote control and scheduling of temperature based on preferences and habits.
- **Smart lighting:** Allows remote control and scheduling of lighting, as well as control through voice commands and color/brightness adjustments.

- **Smart security:** Includes cameras, door locks, and sensors that enable remote monitoring and control of access to the home, with alerts for suspicious activity.
- **Smart appliances:** Includes devices such as refrigerators, ovens, and washing machines that can be controlled remotely, with alerts for issues and scheduling based on preferences.
- **Smart speakers and voice assistants:** Includes Alexa, Amazon Echo, and Google Home that allow control of various aspects of the home through voice commands and provide services such as music, news, and weather.

The main functionalities of a smart home include remote control and automation, scheduling and adjusting settings, monitoring and alerting, and voice control. The specific components and functionalities may vary depending on the needs and preferences of the residents, as it can be seen in Figure 1.4 illustrated by characters. The temperature of a room and the brightness can be adjusted automatically from the communication between devices with sensors (as illustrated in part "a" of Figure 1.4) and the air conditioning or with a virtual assistant as Alexa, for example. Selected background music can start playing when you get home and even the carbon dioxide levels in a room can be monitored and regulated to avoid health problems arising from this (shown in part "b" of Figure 1.4). In domestic environments, in the future, it may be possible to make video calls through TV monitors (in part "c" of Figure 1.4). Currently, there are already solutions for making toast with the weather forecast for that day (part "d" of Figure 1.4) and smart showers (part "e" of Figure 1.4) that play music, regulate the temperature to your appreciation, change the brightness of the bathroom, and even remind users to conserve water with alerts of potential water waste due to a long bath. Devices in smart homes can be used to monitor the elderly (part "f" of Figure 1.4) and detect falls that can be fatal for them. With this,

FIGURE 1.4 A funny illustration of the elements in a smart home.

the detectors can communicate via the internet with other devices and systems and even call the emergency to deal with accidents that could be fatal, as shown in part "g" of Figure 1.4. Smart homes will be addressed again in Section 2.4 in Chapter 2.

1.3.3   SMART HOSPITALS AND SMART CARE

Chapter 8 delves deeper into the concept of healthy cities and it will discuss further the smart hospitals, which are facilities and services that employ technology to enhance patient care and decrease costs. Smart homes can also play a role in smart care, such as through the use of monitoring devices for elderly individuals, which are connected to emergency systems to call for help in case of a fall. A smart hospital, on the other hand, employs technology to improve patient care, decrease costs, and enhance the overall hospital experience. Key components of a smart hospital include:

- **Electronic health records (EHRs):** Digital versions of patients' medical records that can be accessed by authorized personnel from any location, improving coordination of care and reducing administrative work. Some EHRs are recorded using blockchain, a topic discussed in Chapter 6.
- **Remote monitoring:** Smart hospitals use remote monitoring systems to track patients' vital signs and other health information, allowing for quick response to changes in the patient's condition.
- **Robotics and automation:** Smart hospitals utilize robots and automation technology for tasks such as dispensing medication and transporting patients, reducing human error, and increasing efficiency.
- **Telemedicine:** Allows patients to receive medical care remotely, such as through video conferencing with doctors or remote monitoring of vital signs.
- **Decision support systems:** Computer-based systems that provide doctors and nurses with relevant information, such as treatment options and drug interactions, to improve the quality of care.
- **Smart beds and equipment:** Beds and other equipment equipped with sensors and other technology to monitor patients' vital signs, increase patient comfort, and reduce the risk of bedsores.
- **Smart operating rooms:** Use technology to improve the precision and efficiency of surgeries, with the use of robots, real-time monitoring, and other technology.

1.3.4   SMART WASTE MANAGEMENT

Smart waste management uses technology to improve the efficiency and sustainability of waste collection and disposal. It includes several key components, such as:

- **Smart bins:** These are specially designed waste containers that are equipped with sensors and wireless technology, as that displayed in Figure 1.5. They can communicate information about their fill level to waste management companies, so they can be collected and emptied more efficiently. Selective

FIGURE 1.5   Smart bins in the context of a smart campus project.

collection can also be performed in an easier manner, with computational vision and AI helping users to classify where to put each part of the waste. The bin displayed in Figure 1.5 is part of a project named SOFTWAY4IoT (SOFTware-defined gateWAY and fog computing for Internet of Things). Four smart waste bins were placed at the entrance to the faculty building at the Federal University of Goiás, in Goiânia, Brazil, arranged according to the proper separation, that is, one waste bin for each category: glass, paper, metal, and plastic. Each bin has an IoT device for monitoring and managing selective collection, as well as an IP camera. The managers are notified as the bin becomes full so that the discarded material can be recycled (Oliveira Jr et al., 2020).

- **Smart route optimization:** This is a system that uses data from smart bins and other sources to optimize the routes of waste collection vehicles. It helps to reduce the number of miles driven by waste collection vehicles and the amount of fuel consumed, which in turn reduces carbon emissions and costs.
- **Real-time monitoring:** Smart waste management system allows real-time monitoring of the waste collection and disposal process, which

helps waste management companies to identify and address any issues in a timely manner.

- **Recycling and composting:** Smart waste management systems also often include initiatives to increase recycling and composting rates. This can include education campaigns, incentives for recycling, and the use of smart bins that are specifically designed for recycling or composting.
- **Smart landfills:** Smart landfills are equipped with sensors and monitoring systems that can track the levels of waste and emissions, and also track the settlement of the waste and the stability of the landfill.

All these components work together to create a more efficient, sustainable, and cost-effective waste management system that can reduce environmental impact, improve public health, and save money for both municipalities and businesses.

1.3.5 EMERGENCY MANAGEMENT AND RESPONSE SYSTEMS (EMRS)

EMRS utilize technology such as Wireless River Sensors, Telecommunication Gateways, UAVs, VANETs, Meteorological Centers, Fire and Rescue Services, Hospital Centers, Police Departments, Short Message Service Centers, and Social Networks to quickly respond to emergencies and natural disasters. This system is often linked to a National Center for Natural Disaster Monitoring that monitors multiple cities, thousands of sensors, and hydrological sensors to detect potential floods.

An example of a crucial component of EMRS is the Flood Monitoring System (FMS), which can notify possible emergency situations to residents, businesses, and government entities in areas at risk of flooding. FMS is composed of various systems such as smart sensors, gateways, crowd-sourcing systems, drones, and drone bases. These systems work together to emit flood alerts to public authorities, who can then take action to protect the population (Ueyama et al., 2017; Graciano Neto et al., 2017; Graciano Neto et al., 2018a,b; Horita et al., 2017, 2019).

1.3.6 SMART GRIDS FOR SMARTER POWER CONSUMPTION, DISTRIBUTION, AND MANAGEMENT

A smart grid is a modernized electrical system that uses technology to improve efficiency, reliability, and flexibility. It includes various components that work together to optimize energy resources and reduce environmental impact. These include advanced metering infrastructure (AMI) for real-time energy usage data communication, distribution automation (DA) for efficient grid management, energy storage systems (ESS) for storing excess renewable energy, advanced control systems for balancing supply and demand, integration of renewable energy sources such as solar and wind power, and electric vehicle (EV) charging stations to reduce overall load on the grid.

The following box shows how this principle can be associated with architecting and design practices to make a public space smarter in many senses, being part of the landscape and also part of an optimized power distribution system.

SMARTER CONSTRUCTS UNDER ARCHITECTS' AND DESIGNERS' POINT OF VIEW

Smart cities are not only concerned with technology. As it is further discussed in Section 1.7, a smarter city implies smarter decisions, smarter citizen, and also a smarter architecting. This means that a smart city is expected to exhibit optimized structures and forms, with ergonomic characteristics. A public park can have ergonomic and innovative benches, with different forms that could be reused for other purposes, or artificial structures forming vertical gardens that resemble imposing trees, with wide canopies and colored lights at night, covered with plants for sustainability purposes, such as the Supertree Grove in Singapore. A lamppost can also have solar panels to capture and store energy besides having a decoration purpose, with a format close to a palm tree, as depicted in Figure 1.6. Then, at that context, the lamppost has three purposes: (1) illumination, (2) decoration, and (3) sustainability by saving energy. The structure itself was then optimized, and became a smarter solution. Architects and designers can use their creativity to make the urban environment more innovative, beautiful, and optimized.

FIGURE 1.6   An example of smart design in Aparecida de Goiânia, Brazil.

1.3.7 OTHER SMART SYSTEMS

A smart city is composed of several other smart systems. Some of them we will detail in forthcoming chapters; others we briefly outline, as follows.

- **Smart agribusiness:** These agricultural practices use technology to optimize crop yields (agriculture) and related practices, as livestock and poultry farming, maximizing the vegetables, and animal development while reducing as much as possible the environmental impact. Chapter 9 delves into this topic with more details.
- **Smart traffic control systems:** Smart traffic control systems include technologies such as autonomous cars, navigation applications, and smart crosswalks to improve traffic flow, reduce congestion, and enhance safety. This is also linked with mobility in cities and how to plan the urban space. The adoption of simulation technology to plan smart cities infrastructure is detailed in Chapter 4 and mobility is covered in Chapter 7. Smart traffic control systems are also influenced by the next important part of a smart city: emergency management and response systems.
- **Smart campus:** Evidently, in a smart city, the education institutions also are required to be smart. Some initiatives around the world have explored the requirements to build a smart campus (França et al., 2020), that is, a university campus that adopts disruptive technologies to provide better educational experiences and outcomes for its students, staff, and faculty. Chapter 10 is entirely dedicated to dive into the smart education world.
- **Smart water management:** Smart water management uses technology to improve the efficiency, reliability, and sustainability of water distribution and treatment. Their structure includes:
  - **Smart meters:** These are meters that measure water usage and communicate that data to water utilities in real-time. This allows utilities to identify and address leaks and other issues more quickly and also helps customers to monitor their own water usage and identify ways to conserve.
  - **Advanced leak detection:** Smart water management systems include advanced sensors and algorithms to detect leaks in water pipelines and infrastructure. This can help to reduce water loss and costs for water utilities.
  - **Real-time monitoring:** Smart water management systems allow for real-time monitoring of water quality, flow rates, and other important parameters. This helps water utilities to identify and address issues more quickly, and also helps to ensure that water is safe to drink.
  - **Water conservation:** Smart water management systems often include initiatives to increase water conservation, such as education campaigns, incentives for water-efficient appliances, and the use of smart irrigation systems that optimize watering schedules based on weather and other factors.
  - **Advanced water treatment:** Smart water management systems include advanced water treatment technologies that can remove contaminants and other impurities from water more effectively, which can help to improve water quality and reduce the environmental impact of water treatment.

The integration of various components within a smart water management system, such as smart meters, advanced leak detection, real-time monitoring, predictive maintenance, water conservation initiatives, and advanced water treatment technologies, work in unison to create an optimized water management system. This system is characterized by increased efficiency, reliability, and sustainability, which in turn leads to cost reduction, improvement in public health, and protection of the environment.

A SMARTER DEATH

In a smart city, even dying should be a smarter act. Researchers have investigated over the years how to make it real. Maciel and Pereira (2013) investigated the digital legacy, that is, how your image, data, and assets are managed after you die. Passwords, photos, and social network accounts can cause emotional pain to relatives and friends. Then, policies have been thought, established, and adopted in social networks. Under another perspective, cemeteries, and coffins also have their days numbered. Companies have proposed confining bodies in organic coffins so that the body can nourish a tree, and the resulting tree will be a sustainable headstone for that person, and cemeteries would become parks.

1.4   SMART CITIES MAIN TECHNOLOGIES

Smart cities utilize a variety of technologies to improve the efficiency and livability of urban environments. IoT, cloud computing, AI/machine learning, blockchain, digital twins, and system of systems are some of the key technologies that play a vital role in this endeavor.

IoT devices such as sensors and cameras are used to collect data on traffic flow, energy consumption, and environmental conditions. We will discuss further the role of IoT in smart cities in Chapter 2, since IoT comprises the backbone that underpins smart cities.

These data can then be analyzed in the cloud using machine learning algorithms to optimize city services and infrastructure.

Blockchain can be used to secure and manage transactions, such as the sharing of data between city departments and private companies. Chapter 6 discusses the blockchain technology in the context of the smart cities.

Simulation models and digital twins enable cities to create virtual models of their infrastructure, allowing for real-time monitoring, and simulation of potential changes. Former is discussed in Chapter 4, while latter is discussed further in Chapter 5.

Finally, systems-of-systems approach allows for the integration of various technologies and data sources, creating a holistic view of the city, and enabling more effective decision-making. We discuss the systems of systems in Chapter 3.

1.5 SMART CITIES CONCEPTUAL FRAMEWORKS

A conceptual framework for smart cities is a set of principles, concepts, and models that define the main components of a smart city and how they interact and interrelate to create a more sustainable, efficient, and livable urban environment. Examples of key concepts and principles that are often included in a conceptual framework for smart cities include:

- **Integrated and holistic approach:** Smart cities are seen as systems that require an integrated and holistic approach to planning, management, and governance, in order to optimize the use of resources and reduce the environmental impact of urban development.
- **Sustainability (the concept discussed in Section 1.2):** Smart cities are designed to be sustainable, which means they are able to meet the needs of the present without compromising the ability of future generations to meet their own needs. This includes reducing energy consumption, promoting renewable energy sources, and reducing waste and pollution.
- **Livability:** Smart cities are designed to improve the quality of life for residents, by providing access to services, transportation, and amenities that meet the needs of different population groups, such as children, the elderly, and people with disabilities.
- **Resilience:** Smart cities are designed to be resilient to natural disasters and other disruptions, by building in redundancy and back-up systems, and by promoting the use of renewable energy sources and other technologies that can withstand extreme weather events and other disruptions.
- **Smart mobility:** Smart cities are designed to improve the mobility of people and goods, by promoting the use of public transportation, walking, and cycling, and by reducing congestion and pollution.
- **Smart and digital governance:** Smart cities are designed to promote good governance, by involving citizens and stakeholders in the decision-making process, by increasing transparency, and by using data and technology to improve the management and delivery of services. There are several initiatives and forums over the world discussing how to make it real (Reis et al., 2021a,b).

NIST Special Publication 1900-206, *Smart Cities and Communities: A Key Performance Indicators Framework*, is a publication developed by the National Institute of Standards and Technology (NIST) in the United States. The publication provides a framework for measuring the performance of smart cities and communities, by providing a set of key performance indicators (KPIs) that can be used to assess the effectiveness of smart city initiatives. The framework was designed to be flexible and adaptable, so that it can be used in different types of smart cities and communities, and to track progress over time.

Maturity models have also been proposed to evaluate how adherent a city is to the principles of being a smart city. A maturity model is a framework used to evaluate and measure the progress of an organization or project in a specific area. It typically includes a set of stages or levels of maturity, with each stage representing an increase in capabilities and capacity. Maturity models are often used to assess the readiness of an organization or project to implement new technologies or practices, such as in the case of smart cities.

There are several maturity models for smart cities, including:

- **The European Smart Cities and Communities Lighthouse (SCCL) Maturity Model:** This model assesses the maturity of smart cities and communities projects across Europe, focusing on areas such as governance, citizen engagement, and digital services.
- **The Smart City Maturity Model (SCMM):** Developed by the NIST, this model is designed to help cities assess their current level of smart city development and identify areas for improvement.
- **The Smart City Readiness Model (SCRM):** Developed by the Smart Cities Council, this model evaluates the readiness of cities to implement smart city technologies and practices, and provides recommendations for improvement.
- **The Smart Sustainable Cities Maturity Model (SSCMM):** This model was developed by a group of European researchers and focuses on assessing the maturity of smart and sustainable cities in terms of governance, citizen engagement, and sustainable development.

Rankings have also been created to classify how smart a city is: https://www.smart-cities.eu/. Ranking of smart cities refers to the process of evaluating and comparing cities based on their level of development and implementation of smart city technologies and practices. This can include areas such as transportation, energy, communication, and citizen engagement. There are several organizations and institutions that conduct rankings of smart cities. Some examples include:

- **The Smart City Index:** This index, developed by IESE Business School, ranks cities based on their level of smart city development in areas such as mobility, governance, and the environment.
- **The Smart City Index by Juniper Research:** This index ranks cities based on their level of smart city development in areas such as transportation, energy, and citizen engagement.
- **The Smart City Ranking by the Smart City Institute:** This ranking evaluates cities based on their level of smart city development in areas such as transportation, energy, and citizen engagement.
- **The Smart Cities Index by the Centre for Economic and Business Research (CEBR):** This index ranks cities based on their level of smart city development in areas such as transportation, energy, and citizen engagement.

Gartner, a leading research and advisory company, has also published several reports and articles on the topic of smart cities. Gartner also predicts that by 2025, at least 80% of all smart city initiatives will have to be redesigned, due to the lack of clear return on investment (ROI) and a failure to engage citizens (Duncan, 2021).

1.6 EXAMPLES OF SMART CITIES INITIATIVES IN THE WORLD

There are different examples of smart cities initiatives around the world. This section provides a panoramic overview of some of these:

- **Singapore:** Singapore is often cited as one of the most advanced smart cities in the world. It has been used as an example of an *Avant-Garde* city that started to implement the smart city concepts in the end of the 1990s (Mahizhnan, 1999; Lee et al., 2016). The city-state has implemented a wide range of smart city initiatives, including a citywide sensor network to monitor traffic, weather, and air quality, a smart transportation system that uses real-time data to optimize traffic flow, and a smart grid that allows residents to track and manage their energy consumption. Additionally, the government has launched a number of digital services, such as e-payment systems and online government services, to improve the efficiency and accessibility of public services.
- **Barcelona:** The city of Barcelona has been a leader in smart city initiatives in Europe. The city has implemented a wide range of initiatives, including a smart lighting system that adjusts the brightness of streetlights based on the presence of pedestrians and vehicles, a smart waste management system that uses sensors to optimize waste collection, and a smart transportation system that uses real-time data to improve traffic flow and public transportation. Additionally, the city has launched a number of digital services, such as a mobile application that allows residents to report issues and request services, and a platform that allows residents to track the city's budget and spending.
- **Amsterdam:** Amsterdam has been a leader in smart city initiatives in the Netherlands. The city has implemented a wide range of initiatives, including a smart grid that allows residents to track and manage their energy consumption, a smart traffic management system that uses real-time data to optimize traffic flow, and a smart parking system that helps drivers find available parking spaces. Additionally, the city has launched a number of digital services, such as a platform that allows residents to report issues and request services, and a platform that allows residents to track the city's budget and spending.
- **New York City, United States:** New York City has implemented a number of smart city initiatives under its NYCx program, which aims to use technology to address urban challenges. The initiatives include a smart street lighting system that adjusts the brightness of streetlights based on the presence of pedestrians and vehicles, a smart waste management system that uses sensors to optimize waste collection, and a smart traffic management system that uses real-time data to improve traffic flow and public transportation. Additionally, the city has launched a number of digital services, such as a mobile application that allows residents to report issues and request services, and a platform that allows residents to track the city's budget and spending.

- **Toronto, Canada:** Toronto has implemented several smart city initiatives, including a smart transportation system that uses real-time data to optimize traffic flow, a smart lighting system that adjusts the brightness of streetlights based on the presence of pedestrians and vehicles, and a smart waste management system that uses sensors to optimize waste collection. Additionally, the city has launched a number of digital services, such as a mobile application that allows residents to report issues and request services, and a platform that allows residents to track the city's budget and spending.
- The Emerging and Sustainable Cities Initiative (ICES) is a project of the Inter-American Development Bank (IDB) that aims to promote sustainable urban development in Latin America and the Caribbean. The initiative focuses on improving the quality of life for residents of participating cities by addressing issues such as mobility, housing, and the environment. The ICES project is designed to be implemented in stages, with each stage building on the progress made in the previous stage. The goal of the initiative is to create sustainable and livable cities that are able to meet the needs of their residents and contribute to the overall development of the region.

1.7   SOCIAL IMPACT OF SMART CITIES

Smart cities have the potential to bring several benefits for the population, but also an extensive social impact. One important assumption is that, for cities to become smart, their citizens should also become smarter. This means that smart city inhabitants should also be well informed and educated about the technologies and systems that make up a smart city. Public policies should be established to involve all the layers of our society to be integrally benefited from the advances brought by technology. As such, it is important having programs to qualify and widespread knowledge and abilities to interact with smart devices in smart cities, particularly underprivileged neighborhoods and their low-income populations (Kon et al., 2020).

One solution for including underprivileged neighborhoods in smart city solutions is to involve community members in the planning and implementation process. This can be done through community meetings and workshops, where residents can provide input on what smart city solutions would be most beneficial for their neighborhood and how they can be implemented. Another solution is to ensure that smart city solutions are designed to address the specific needs and challenges faced by underprivileged communities, such as improving access to education and job training, reducing crime, and increasing access to transportation and healthcare.

One of the challenges to including underprivileged neighborhoods in smart city solutions is that they may not have the same level of access to technology and the internet as more affluent neighborhoods. This can make it difficult to implement solutions that rely on connectivity, such as smart lighting or traffic control systems. Additionally, some residents may not have the skills or resources to fully utilize smart city technologies, which can limit their ability to take advantage of the benefits they provide.

However, it is imperative that smart cities solutions are planned to involve all the tiers of our society, being inclusive and accessible for all citizens, regardless of their socioeconomic status or location. By involving community members in the planning and implementation process, and by designing solutions to address the specific needs and challenges faced by underprivileged neighborhoods, it is possible to create smart cities that work for everyone.

1.8 FINAL REMARKS

This chapter acted as an introductory to the topic of smart cities, providing definitions and discussing the key components, technologies, and frameworks involved.

Upon examining what constitutes a smart city, it becomes clear that achieving a more intelligent, integrated, and innovative urban environment requires a comprehensive and systemic approach, as well as the seamless integration of various actors and sectors. To accomplish this, it is crucial to not only focus on technological advancements but also to innovate in management, planning, governance models, and public policy development. Adopting a holistic approach is essential in effectively transforming the urban landscape. Other chapters address the challenges facing several crucial dimensions of smart cities. The next chapter covers IoT, a disruptive technology that enabled smart cities to come true.

REFERENCES

Alampalli, S., & Pardo, T. (2014). A study of complex systems developed through public private partnerships. In Proceedings of the 8th International Conference on Theory and Practice of Electronic Governance – ICEGOV '14, pp. 442–445. https://doi.org/10. 1145/2691195.2691212

Alexander, C. (1977). A Pattern Language: Towns, Buildings, Construction. Oxford University Press.

Anderson, C. (2008). The end of theory: The data deluge makes the scientific method obsolete. Wired Magazine, 16(7), https://www.wired.com/2008/06/pb-theory/

Arruda, M. F., & Bulcão-Neto, R. F. (2019). Toward a lightweight ontology for privacy protection in IoT. In Proceedings of the 34th ACM/SIGAPP Symposium on Applied Computing (SAC '19) (pp. 880–888.) Association for Computing Machinery, New York, NY. https://doi.org/10.1145/3297280.3297367

Asiag, J. J. (2020). 8 Smart Cities Lead the Way in Advanced Intelligent Transportation Systems. Otonomo. https://otonomo.io/blog/smart-cities-intelligent-transportation-systems/

Boscarioli, C., Araujo, R. M., & Maciel, R. S. P. (2017). I GrandSI-BR – Grand Research Challenges in Information Systems in Brazil 2016–2026. Special Committee on Information Systems (CE-SI). Brazilian Computer Society (SBC). 184p ISBN: [978-85-7669-384-0].

Burns, R., Stevens, J. A., & Lee, R. (2016). The direct costs of fatal and non-fatal falls among older adults – United States. Journal of Safety Research, 58, 99–103.

Cavalcante, E., Cacho, N., Lopes, F., & Batista, T. (2017). Challenges to the development of smart city systems: A system-of-systems view. In Proceedings of the Brazilian Symposium on Software Engineering, 244–249.

Chaudhuri, S., Thompson, H., & Demiris, G. (2014). Fall detection devices and their use with older adults: A systematic review. Journal of Geriatric Physical Therapy, 920(37), 178.

Costa, F. M., Morris, K. A., Kon, F., & Clarke, P. J. (2017). Model-driven domain-specific middleware. In Proceedings of the International Conference on Distributed Computing Systems 2017 (pp. 1961–1971). Atlanta, GA.

Cui, L., Xie, G., Qu, Y., Gao, L., & Yang, Y. (2018). Security and privacy in smart cities: Challenges and opportunities. IEEE Access, 6, 46134–46145. https://doi.org/10.1109/ACCESS.2018.2853985

da Silva, W. M., Alvaro, A., Tomas, G. H. R. P., Afonso, R. A., Dias, K. L., & Garcia, V. C. (2013). Smart cities software architectures: A survey. In Proceedings of the 28th Annual ACM Symposium on Applied Computing (SAC '13) (pp. 1722–1727). Association for Computing Machinery, New York, NY USA. https://doi.org/10.1145/2480362.2480688

Deguchi, A. (2020). From smart city to society 5.0. Society, 5, 43–65.

Delécolle, A., Lima, R. S., Graciano Neto, V. V., & Buisson, J. (2020). Architectural strategy to enhance the availability quality attribute in system-of-systems architectures: A case study. In Proceedings of the System of Systems Engineering Conference 2020 (pp. 93–98). Budapest, Hungary.

Donoso, P., Martínez, F., & Zegras, C. (2006). The Kyoto Protocol and sustainable cities: Potential use of clean-development mechanism in structuring cities for carbon-efficient transportation. Transportation Research Record, 1983(1), 158–166.

Duncan, A. D. (2021). "Over 100 Data and Analytics Predictions Through 2025," Gartner, Inc. https://mpost.io/wp-content/uploads/Gartner-100-data-analytics-predictions-2025.pdf

França, Í., Araújo, N., Gomes, A., Cacho, N., Lopes, F., Lima, J., & Adachi, E. (2020). SIGOc: A smart campus platform to improve public safety. In 2020 IEEE 6th World Forum on Internet of Things (WF-IoT) (pp. 1–6). IEEE.

Giannoutakis, K. M., Spanopoulos-Karalexidis, M., Filelis Papadopoulos, C. K., & Tzovaras D. (2020). Next generation cloud architectures. In: Lynn T., Mooney J., Lee B., Endo P. (eds) The Cloud-to-Thing Continuum. Palgrave Studies in Digital Business & Enabling Technologies. Palgrave Macmillan, Cham. https://doi.org/10.1007/978-3-030-41110-7_2

Giffinger, R., & Gudrun, H. (2010). Smart cities ranking: An effective instrument for the positioning of the cities? ACE: Architecture, City and Environment, 4(12), 7–26.

Graciano Neto, V. V., Barros Paes, C. E., Garcés, L., Guessi, M., Manzano, W., Oquendo, F., & Nakagawa, E. Y. (2017). Stimuli-SoS: A model-based approach to derive stimuli generators for simulations of systems-of-systems software architectures. Journal of the Brazilian Computer Society, 23(1), 1–22.

Graciano Neto, V. V., Horita, F. E. A., Santos, R. P. D., Viana, D., Kassab, M., Manzano, W. A. E., & Nakagawa, E. Y. (2019). Sob (save our budget): A simulation-based method for prediction of acquisition costs of constituents of a system-of-systems. Revista Brasileira de Sistemas de Informação – iSys, 12(4), 6–35.

Graciano Neto, V. V., Manzano, W., Kassab, M., & Nakagawa, E. Y. (2018a). Model-based engineering & simulation of software-intensive systems-of-systems: Experience report and lessons learned. In Proceedings of the European Conference on Software Architecture (Companion) (27, 1–27:7). Madrid, Spain.

Graciano Neto, V. V., Manzano, W., Rohling, A. J., Ferreira, M. G. V., Volpato, T., & Nakagawa, E. Y. (2018b). Externalizing patterns for simulations in software engineering of systems-of-systems. In Proceedings of the 33rd Annual ACM Symposium on Applied Computing(pp. 1687–1694).

Graciano Neto, V. V., Santos, R. P., Viana, D., & Araujo, R. (2020). Towards a conceptual model to understand software ecosystems emerging from systems-of-information systems. In: Santos R., Maciel C., Viterbo J. (eds) Software Ecosystems, Sustainability and Human Values in the Social Web. WAIHCWS 2017, WAIHCWS 2018. Communications in Computer and Information Science, vol 1081. Springer, Cham. https://doi.org/10.1007/978-3-030-46130-0_1

Graciano Neto, V. V. (2018). A simulation-driven model-based approach for designing software-intensive systems-of-systems architectures. (Une approche dirigée par les simulations à base de modèles pour concevoir les architectures de systèmes-des-systèmes à logiciel prépondérant). PhD Thesis. University of Southern Brittany, Vannes, Morbihan, France.

Gutierrez-Madroñal, L., Blunda, L. L., Wagner, M., & Medina-Bulo, I. (2019). Test event generation for a fall-detection IoT system. IEEE Internet of Things Journal, 6, 6642–6651.

Harrison, C., Eckman, B., Hamilton, R., Hartswick, P., Kalagnanam, J., Paraszczak, J., & Williams, P. (2010). Foundations for smarter cities. IBM Journal of Research and Development, 54(4), 1–16.

Horita, F. E., de Albuquerque, J. P., Marchezini, V., & Mendiondo, E. M. (2017). Bridging the gap between decision-making and emerging big data sources: An application of a model-based framework to disaster management in Brazil. Decision Support Systems, 97, 12–22.

Horita, F. E., Rhodes, D. H., Inocêncio, T. J., & Gonzales, G. R. (2019). Building a conceptual architecture and stakeholder map of a system-of-systems for disaster monitoring and early-warning: A case study in Brazil. In Proceedings of the XV Brazilian Symposium on Information Systems (pp. 1–8).

IEEE P2413. (2019a). Standard for an Architectural Framework for the Internet of Things (IoT). IEEE Standards Association, 21 May. Accessed November 2020. https://standards.ieee.org/standard/2413-2019.html

IEEE P2413.1. (2019b). Standard for a Reference Architecture for Smart City (RASC). IEEE Standard Association. Accessed November 2020. https://standards.ieee.org/project/2413_1.html

ISO. (2014). ISO 37120 – Sustainable development of communities – Indicators for city services and quality of life.

ISO/IEC. (2017). ISO/IEC 19941:2017 – Information technology – Cloud computing – Interoperability and portability.

ITU-T. (2012). Recommendation Y.2060: Overview of the Internet of Things. http://handle.itu.int/11.1002/1000/11559

ITU-T Focus Group on Smart Sustainable Cities. (2014). Smart Sustainable Cities: An Analysis of Definitions. Focus Group Technical Report, Geneva, Switzerland, Tech. Rep. FG-SSC-10/2014.

Khanna, A., & Kaur, S (2020). Internet of Things (IoT), applications and challenges: A comprehensive review. Wireless Personal Communications, 114, 1687–1762. https://doi.org/10.1007/s11277-020-07446-4

Kon, F., Braghetto, K., Santana, E. Z., Speicys, R., & Guerra, J. G. (2020). Toward smart and sustainable cities. Communications of the ACM, 63(11), 51–52.

Kon, F., & Santana, E. F. Z. (2016). Cidades Inteligentes: Conceitos, plataformas e desafios. Jornadas de atualização em informática, 17. Brazilian Computer Society.

Kruchten, P., Obbink, H., & Stafford, J. (2006). The past, present, and future for software architecture. IEEE Software, 23(2), 22–30.

Lee, S. K., Kwon, H. R., Cho, H., Kim, J., & Lee, D. (2016). International Case Studies of Smart Cities: Singapore, Republic of Singapore. Inter-American Development Bank. https://publications.iadb.org/en/international-case-studies-smart-cities-singapore-republic-singapore

Lynn, T., Endo, P. T., Ribeiro, A. M. N. C., Barbosa, G. B. N., & Rosati, P. (2020). The Internet of Things: Definitions, key concepts, and reference architectures. In: Lynn T., Mooney J., Lee B., Endo P. (eds) The Cloud-to-Thing Continuum. Palgrave Studies in Digital Business & Enabling Technologies. Palgrave Macmillan, Cham. https://doi.org/10.1007/978-3-030-41110-7_1

Maciel, R. S. P., David, J. M. N., Claro, D. B., & Braga, R. (2017). Full Interoperability: Challenges and Opportunities for Future Information Systems. Grand Challenges in Information Systems for the next 10 years (2016-2026), SBC, 107–118.

Maciel, C., & Pereira, V. (2013). Digital Legacy and Interaction. Springer, Heidelberg.

Mahizhnan, A. (1999). Smart cities: The Singapore case. Cities, 16(1), 13–18.

Manzano, W., Graciano Neto, V. V., Nakagawa, E. Y. (2020). Dynamic-SoS: An approach for the simulation of systems-of-systems dynamic architectures. The Computer Journal, 63(5), 709–731.

Maranhão, G. M., & Bulcão-Neto, R. (2016). A semantic filtering mechanism geared towards context dissemination in ubiquitous environments. Journal of Universal Computer Science, 22, 1123–1147.

Mauldin, T. R., Canby, M. E., Metsis, V., Ngu, A. H., & Rivera, C. C. (2018). Smartfall: A smartwatch-based fall detection system using deep learning. Sensors, 18, 3363.

Mohanty, S. P., Choppali, U., & Kougianos, E. (2016). Everything you wanted to know about smart cities: The internet of things is the backbone. IEEE Consumer Electronics Magazine, 5(3), 60–70.

Morin, B., Harrand, N., & Fleurey, F. (2017). Model-based software engineering to Tame the IoT jungle. IEEE Software, 34(1), 30–36. https://doi.org/10.1109/MS.2017.11

Motta, R., de Oliveira, K., & Travassos, G. (2019). On challenges in engineering IoT software systems. Journal of Software Engineering Research and Development, 7, 5:1–5:20. https://doi.org/10.5753/jserd.2019.15

Mulley G. (2001). Falls in older people. Journal of the Royal Society of Medicine, 94(4), 202.

Oliveira Jr, A., Cardoso, K., Sousa, F., & Moreira, W. (2020). A lightweight slice-based quality of service manager for IoT. IoT, 1(1), 4.

Perera, C., Barhamgi, M., Bandara, A. K., Ajmal, M., Price, B., & Nuseibeh, B. (2020). Designing privacy-aware internet of things applications. Information Sciences, 512, 238–257, ISSN 0020-0255. https://doi.org/10.1016/j.ins.2019.09.061

Perera, C., Barhamgi, M., De, S., Baarslag, T., Vecchio, M., & Choo, K. R. (2018). Designing the sensing as a service ecosystem for the internet of things. IEEE Internet of Things Magazine, 1(2), 18–23. https://doi.org/10.1109/IOTM.2019.1800023

Perera, C., Zaslavsky, A., Christen, P., & Georgakopoulos, D. (2014). Context aware computing for the internet of things: A survey. IEEE Communications Surveys & Tutorials, 16(1), 414–454, First Quarter 2014. https://doi.org/10.1109/SURV.2013.042313.00197

Perry, D. E., & Wolf, A. L. (1992). Foundations for the study of software architecture. ACM SIGSOFT Software Engineering Notes, 17(4), 40–52.

Projetos, F. G. V. (2014). Cidades inteligentes e mobilidade urbana. Cadernos FGV Projetos, Rio de Janeiro, (24).

Reis, L. C. D., Bernardini, F. C., Leal Ferreira, S. B., & Cappelli, C. (2021a). Exploring the challenges of ICT governance in Brazilian smart cities. In Proceedings of the 14th International Conference on Theory and Practice of Electronic Governance (pp. 429–435). ACM.

Reis, L. C. D., Bernardini, F. C., Leal Ferreira, S. B., & Cappelli, C. (2021b). ICT governance in Brazilian smart cities: An integrative approach in the context of digital transformation. In DG.O 2021: The 22nd Annual International Conference on Digital Government Research(pp. 302–316). ACM, Omaha, NE.

Rocha, V., Alves, L., Graciano Neto, V. V., & Kassab, M. (2019). A Review on the Adoption of Agile Methods in the Technology Development for Smart Cities. In WORKSHOP BRASILEIRO DE CIDADES INTELIGENTES (WBCI), 2., 2019, Belém. Anais […]. Porto Alegre: Sociedade Brasileira de Computação. https://doi.org/10.5753/wbci.2019.6748

Santana, E. F. Z., Chaves, A. P., Gerosa, M. A., Kon, F., & Milojicic, D. S. (2018). Software platforms for smart cities: Concepts, requirements, challenges, and a unified reference architecture. ACM Computing Surveys, 50(6), Article 78, 37. https://doi.org/10.1145/3124391

Santos, G. L., Monteiro, K. H. C., & Endo, P. T. (2020). Living at the edge? Optimizing availability in IoT. In: Lynn T., Mooney J., Lee B., Endo P. (eds) The Cloud-to-Thing Continuum. Palgrave Studies in Digital Business & Enabling Technologies. Palgrave Macmillan, Cham. https://doi.org/10.1007/978-3-030-41110-7_5

Shaw, M., & Clements, P. (2006). The golden age of software architecture. IEEE Software, 23(2), 31–39.

Sheng, X., Tang, J., Xiao, X., & Xue, G. (2013). Sensing as a service: Challenges, solutions and future directions. IEEE Sensors Journal, 13(10), 3733–3741. https://doi.org/10.1109/JSEN.2013.2262677

Shrivastava, R., & Pandey, M. (2020). Real time fall detection in fog computing scenario. Cluster Computing, 23, 2861–2870. https://doi.org/10.1007/s10586-020-03051-z

Spinellis, D. (2017). Software-engineering the internet of things. IEEE Software, 34(1), 4–6. https://doi.org/10.1109/MS.2017.15

Stevens, J. A., Mahoney, J. E., & Ehrenreich, H. (2014). Circumstances and outcomes of falls among high risk community-dwelling older adults. Injury Epidemiology, 1(1), 5. https://doi.org/10.1186/2197-1714-1-5

Tanna, R. D., Kumar, K. S. V., & Karthika, S. (2017). Analytics as a service for beginners. In Proceedings of the 2017 International Conference on Computational Intelligence in Data Science (ICCIDS), Chennai, 1–6. https://doi.org/10.1109/ICCIDS.2017.8272659

Taylor, C., Siebold, A., & Nowzari, C. (2020). On the effects of minimally invasive collision avoidance on an emergent behavior. In: Dorigo M., et al. (eds) Swarm Intelligence. ANTS 2020. Lecture Notes in Computer Science, vol 12421. Springer, Cham. https://doi.org/10.1007/978-3-030-60376-2_27

Ueyama, J., Faiçal, B. S., Mano, L. Y., Bayer, G., Pessin, G., & Gomes, P. H. (2017). Enhancing reliability in wireless sensor networks for adaptive river monitoring systems: Reflections on their long-term deployment in Brazil. Computers, Environment and Urban Systems, 65, 41–52.

UNFCC. (2009). Kyoto protocol reference manual on accounting of emissions and assigned amount. eSocialSciences. https://unfccc.int/sites/default/files/08_unfccc_kp_ref_manual.pdf

United Nations, Department of Economic and Social Affairs, Population Division. (2019). World Urbanization Prospects: The 2018 Revision (ST/ESA/SER.A/420). New York, United Nations. https://population.un.org/wup/Publications/

United Nations, Department of Economic and Social Affairs, Population Division. (2022). 'World Population Prospects 2022: Summary of Results'. Tech Report UN DESA/POP/2022/TR/NO. 3. https://population.un.org/wpp/Publications/

Weyrich, M., & Ebert, C. (2016). Reference architectures for the internet of things. IEEE Software, 33(1), 112–116. https://doi.org/10.1109/MS.2016.20

Yin, C., Xiong, Z., Chen, H., Wang, J., Cooper, D., & David, B. (2015). A literature survey on smart cities. Science China Information Sciences, 58, 1–18.

Zambonelli, F. (2017). Key abstractions for IoT-oriented software engineering. IEEE Software, 34(1), 38–45. https://doi.org/10.1109/MS.2017.3

2 Smart Cities and the Internet of Things
A Synergetic Partnership

2.1 INTRODUCTION TO IoT

Are you ready to get "things" moving? Because we are about to dive into the wild world of the Internet of Things (IoT), where your toaster can talk to your car, your watch can chat with your fridge, and your bathroom scale can have a heated debate with your coffee maker (we do not recommend getting involved in that one). But don't worry, unlike the "Thing" in Marvel's Fantastic Four, the IoT doesn't involve giant gravel monsters dancing hand in hand (although that would be quite a sight!). So, what exactly is this "thing" we're talking about? Let's find out!

The IoT is widely considered to be the driving force behind the development of smart cities (Alampalli and Pardo, 2014). This technology connects a network of physical objects so-called "things," such as electronics, sensors, actuators, and other processing elements, allowing them to collect data from the environment, control other devices, exchange and process data, and interact with users.

By understanding the theoretical foundations of IoT and exploring the potential synergies between IoT and smart cities, we can better understand the future evolution of our cities. This chapter offers a comprehensive overview of IoT and its role in smart cities.

2.2 IoT DEFINED

The term "Internet of Things" has been around since 1999. Kevin Ashton, co-founder of MIT's Auto-ID Labs, coined the term IoT to refer to a combination of real and virtual information technology that are connected through automatic identification technologies (e.g., Radio Frequency IDentification [RFID]), location systems (e.g., global positioning system [GPS]), sensors (e.g., temperature sensor), and actuators (e.g., electric motor). There are many types of sensors, such as RFID tags, GPS, and accelerometers. "Things" could be watches, cameras, social media feeds, and cyber and physical devices – anything that can produce data. The things can be wired or wireless – although most are wireless due to the scale of IoT and limitations of wired infrastructure.

Gartner defines IoT as a network of physical objects (things) that contain embedded technology to communicate and sense or interact with the external environment. The connection of assets, processes, and personnel enables the capture of data and events from which a company can learn behavior and usage, react with preventive action, or augment or transform business processes. The IoT is a foundational

DOI: 10.1201/9781003348542-2

capability for the creation of a digital business (Hung, 2017). Then, a thing could be any physical device embedded with software that can sense, process, and act on environment, and connect to a network.

A GRANDFATHER AND A GRANDMOTHER FOR IoT

IoT was popularized with RFID technology. However, ancient technologies were already crawling to allow IoT run. Examples of that are the Automated Teller Machine (ATM) and Beverage Vending Machines. The former was a cash dispenser invented in the early 1960s embedded with software and hardware so that it could correctly provide a withdraw if the client had balance for that. The latter was a Coke machine at Carnegie Mellon University (CMU). CMU Computer Science Department students created microswitches to put inside the Coke machine to sense how many bottles were available. The software they created for the departmental computer also tracked how recently the bottles were loaded, so students and faculty could ping the computer for information about the availability of soft drinks. CMU was part of ARPANET (the first network between universities in US that originated, later, what we know as Internet) at the time, which made the machine to be connected. The connected nature of that vending machine and the nature of the service it provided made it to be considered an early example of an IoT device. And the nature of a machine with sensors and dispensing cash also can make ATM as one of the grandparents for modern IoT as we know.

Read more at: https://www.machinedesign.com/automation-iiot/article/21836968/iot-started-with-a-vending-machine

The National Institute of Standards and Technology (NIST), Draft Special Publication (800-183), describes the building blocks that "govern the operation, trustworthiness, and lifecycle of IoT" (Voas et al., 2018). NIST is an agency of the U.S. Department of Commerce that is responsible for promoting innovation and industrial competitiveness by advancing measurement science, standards, and technology in ways that enhance economic security and improve our life quality. In order to design and implement IoT innovations, it is extremely important to understand the underlying and foundational science behind IoT; thus, NIST described five primitives for IoT that can be seen as "the anatomy of the thing" (shown in Figure 2.1) and discussed, as follows:

1. Sensor is "an electronic utility (e.g., cameras and microphones) that measures physical properties such as sound, weight, humidity, temperature, and acceleration." Properties of a sensor could be the transmission of data (e.g., RFID), Internet access, and/or be able to output data based on specific events.
2. A communication channel is "a medium by which data are transmitted (e.g., physical via universal serial bus, wireless, wired, and verbal)."

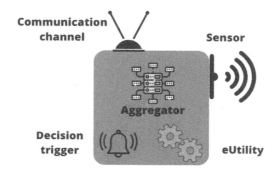

FIGURE 2.1 The anatomy of "the thing."

3. Aggregator is "a software implementation based on mathematical function(s) that transforms groups of raw data into intermediate, aggregated data. Raw data can come from any source." Aggregators have two actors for consolidating large volumes of data into lesser amounts:
   a. Cluster is "an abstract grouping of sensors (along with the data they output) that can appear and disappear instantaneously."
   b. Weight is "the degree to which a particular sensor's data will impact an aggregator's computation."
4. Decision trigger "creates the final result(s) needed to satisfy the purpose, specification, and requirements of a specific IoT." A decision trigger is a conditional expression that triggers an action and abstractly defines the end purpose of an IoT. A decision trigger's outputs can control actuators and transactions.
5. External utility (eUtility) is "a hardware product, software, or service, which executes processes or feeds data into the overall dataflow of the IoT."

What do the "things" do? They capture data and communicate/transmit the data to algorithms (software) that may trigger a decision or perform some action (e.g., an actuator). IoT has enabled many new applications, helped to expand domains such as big data, business and data analytics, and artificial intelligence (AI). Most importantly, the IoT has an impact on quality of life for many.

IoT applications typically involve a general data lifecycle that includes data acquisition, modeling, reasoning, and dissemination (Perera et al., 2014; Maranhão and Bulcão-Neto, 2016), as explained in Figure 2.2. Data are collected from physical or virtual sensors (acquisition), and then processed and stored in an internal, meaningful representation (modeling), such as attribute–value pairs. Data preprocessing techniques may be applied to the modeled data before inference processes are performed, which use decision models to make decisions based on the data (reasoning). Afterwards, actuators carry out the decided actions (or the inference results trigger event notifications) to a third-party system (dissemination).

It is worth noting that depending on the networking, processing, and storage capabilities of a given component, these workflow steps may occur at different

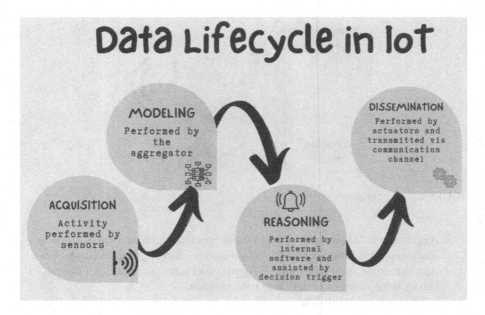

FIGURE 2.2   Data lifecycle in IoT.

components of an IoT system (Lynn et al., 2020). They can take place in the cloud (centralized), on a device, or in an intermediate layer between the devices and the cloud (the fog).

IoT IN THE SUPPLY CHAIN

A supply chain is a network of individuals and companies who are involved in creating a product and delivering it to the consumer. For instance, when you eat meat or vegetables, they were (i) produced by farmers, (ii) transported until a distribution center, (iii) sold to a restaurant or supermarket, and (iv) bought by you. The same rationale is used to industrialized products, when raw material is produced in farms and manufactured in facilities until arriving at the market for consume. IoT has been adopted in supply chains, particularly relying on RFID, as a way to track the sustainability, quality, and animal welfare in production. Organic food receives unique identifiers that can be read by sensors and the final consumer can track its origin and provenance. Moreover, blockchain (a topic covered in Chapter 6) also can be involved to record the entire journey of the food until it reaches your table.

Cloud computing is central to most IoT systems, including smart cities. Cloud is the allusive term used to denote the usage of external servers to store data rather than a local machine. Dropbox is an instance of a cloud. Since files are in the cloud, you can access everywhere, not depending on a physical local machine. Cloud services for smart cities provide services such as sensor data collection, storage, access, analytics, and actuation support, in addition to administrative functions, such as device and user account management and reporting capabilities.

Fog computing, on the other hand, has become a common paradigm in IoT as well. It consists of the idea of preprocessing data before sending it to the central servers, complementing cloud computing by providing cloud services closer to the users (Giannoutakis et al., 2020), and reducing communication latency and data transfer costs to the cloud, which enables real-time applications and services.

If the reader is interested in exploring the definitions of IoT further, we recommend the book *What Every Engineer Should Know About Internet of Things* by DeFranco and Kassab for further reading.

2.3 IoT ECOSYSTEMS

The IoT is becoming increasingly prevalent in many aspects of our daily lives.

According to a report by Gartner, there were an estimated 20 billion IoT devices in use worldwide in 2020, and this number is expected to reach 50 billion by 2030 (Hung, 2017). This growth is driven by factors such as the decreasing cost of IoT devices, small size of technology such as RFID sensors, microelectromechanical devices, Wi-Fi routers, and microcontrollers, as well as the widespread use of mobile devices.

IoT ecosystems can be found (and will appear) in virtually every important application domain including critical infrastructure, entertainment, and every form of pedestrian convenience in the public space and in the home. According to a report by MarketsandMarkets (MarketsandMarkets, 2022), the manufacturing industry is expected to have the largest share of the IoT market, with an estimated market size of $87.9 billion by 2026. In healthcare, IoT devices are used in telemedicine, remote monitoring, and medical equipment management. The global healthcare IoT market is expected to reach $26 billion by 2028. In transportation, IoT technologies are used in areas such as intelligent transportation systems, connected cars, and fleet management.

IoT has also made an enormous impact on our *critical* infrastructure systems. When the "things" are networked *physical* systems, they are called a cyber-physical system (CPS). Critical infrastructure systems fall in the CPS category. A system is deemed critical when the assets, systems, and networks are considered so vital that if compromised, would have a debilitating effect on safety, security, economic security, environmental integrity, and/or public health or safety. IoT allows these physical devices to connect, network, exchange data, and provide real-time decision support. The Cybersecurity and Infrastructure Security Agency (CISA – https://www.cisa.gov/) provides the names and defines the 16 critical infrastructure sectors as follows:

- **Chemical sector:** The part of the economy that manufactures, stores, uses, and transports potentially dangerous chemicals.
- **Commercial facilities sector:** A diverse range of sites that draw large crowds of people for shopping, business, entertainment, or lodging.
- **Communications sector:** The underlying operations for businesses, public safety, and government.

- **Critical manufacturing sector:** Primary metals, machinery, electrical equipment, appliance, transportation, and component manufacturing.
- **Dams sector:** Navigation locks, levees, hurricane barriers, water retention facilities, etc.
- **Defense industrial base sector:** The maintenance of military weapons systems, subsystems, and components or parts to meet US military requirements.
- **Emergency services sector:** Agencies that prevent/prepare/respond/ recover a range of emergency services.
- **Energy sector:** The energy supply.
- Financial services sector: Depository institutions, providers of investment products, insurance companies, other credit and financing organizations, and the providers of the critical financial utilities and services.
- **Food and agriculture sector:** Food manufacturing/processing/storage facilities – also has critical dependency on other critical infrastructure (water, transportation, energy, and chemical).
- **Government facilities sector:** A wide variety of US-owned facilities.
- Healthcare and public health sector: Public and private healthcare organizations that respond and provide recovery across all other sectors in the event of a natural or man-made disaster.
- **Information technology sector:** The virtual and distributed functions that produce and provide hardware, software, and information technology systems and services, and the Internet.
- **Nuclear reactors, materials, and waste sector:** The power reactors that provide electricity to millions of Americans, to the medical isotopes used to treat cancer patients, the nuclear reactors, materials, and waste sector.
- **Transportation systems sector:** This sector includes aviation, highway, maritime, mass transit, passenger rail, pipeline, freight, and postal.
- Water and wastewater systems sector: Public drinking water and wastewater treatment systems.

2.4 MOTIVATIONAL EXAMPLE

Smart homes are a key component of smart cities and serve as an excellent example of how IoT and smart systems are closely intertwined. In our scenario, we imagine a smart home owned by an older adult who requires a specific service – fall detection and notification. Falls are a significant issue for older adults, as they are one of the leading causes of injuries and deaths among this population (Burns et al., 2016; Bulcão-Neto et al., 2023). To address this, the smart home environment is equipped with IoT technologies that monitor and detect falls and notify emergency services accessible to citizens in the city (Chaudhuri et al., 2014; Gutierrez-Madroñal et al., 2019).

In this scenario (illustrated in Figure 2.3), the elderly person can be monitored by both wearable and non-wearable devices. Wearable devices, such as smartphones and smartwatches, can be used in their everyday life (Mauldin et al., 2018). However, when the person is in the bathroom, non-wearable devices such as conventional

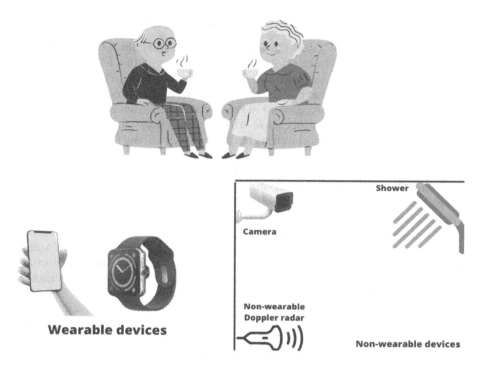

FIGURE 2.3   Wearable and non-wearable devices.

cameras, depth cameras, and Doppler radar can also be used to instantly detect potential falls (Anishchenko et al., 2019). Although a small percentage of falls occur in the bathroom, these are at least twice as likely to result in injury compared to falls in other parts of the home (Mulley, 2001; Stevens et al., 2014). Furthermore, many of these falls go undetected as people often remove their wearable devices when showering, for example.

If a fall is detected, the smart home environment will notify emergency services so that health professionals can assist the elderly and send an ambulance to remove them as soon as possible. The smart home system then communicates with the smart traffic monitoring system to find the fastest and shortest route for the ambulance to reach the elderly's home. Meanwhile, the traffic monitoring system monitors the ambulance's route and responds to any potential traffic issues. Once the ambulance reaches the destination, the elderly adult is then transported to the nearest and best-equipped hospital for emergency treatment and examinations. The destination is a smart hospital that notifies the elderly's emergency contacts as soon as they are admitted, while a wireless body sensor network continuously monitors the patient's vital signs, alerting in-hospital health professionals whenever their health conditions worsen.

Using this example, in the following sections, we will delve into the theoretical and technological foundations that converges IoT and smart cities, explore the anatomy of a smart city from an IoT perspective, and discuss the challenges that remain in the adoption of IoT for smart cities.

2.5   THE ANATOMY OF A SMART CITY UNDER IoT PERSPECTIVE

In the realm of smart cities, the IoT reigns supreme, providing the foundation for the various services on offer. In fact, IoT is a recurrent theme in this field, sharing key characteristics with smart cities, such as complexity, the fact that everything in a smart city is a service, the importance of the spatial and temporal dimensions, and the role of intelligence and big data. As detailed in Perera et al.'s work from 2014 and 2018, the architecture of IoT also plays a critical role in the success of smart cities. We will explore in this section each of these shared characteristics between IoT and the smart cities.

First, both IoT and smart cities are considered as complex systems with many independent and interdependent parts, which interact in non-linear ways (i.e., the behaviors cannot be expressed as the direct conjunction of the activity of individual components) and have interdependencies that are difficult to describe, predict, and design (Alampalli and Pardo, 2014). The IoT includes many physical objects with software, hardware, sensors, and actuators that interact autonomously and are highly dependent on their capabilities, such as storage, processing, and memory (Khanna and Kaur, 2020). Smart cities are composed of several communicating systems that form large networks that can be hard to manage, thus characterizing what is understood as a complex system.

Second, a second shared characteristic is that everything can be regarded as a service. IoT may demand a tremendous amount of infrastructure (e.g., sensing resources) to ease a smart city's deployment (Sheng et al., 2013). Highly interrelated with IoT, the cloud computing paradigm has already shown that consuming resources as a service (platform, infrastructure, or software) is highly efficient, scalable, and easier to use. Thus, this is actually an emerging but already recurrent paradigm that structures everything as a service, with functionalities (e.g., sensing and analytics) that are offered to be invoked by other systems and applications via well-defined interfaces (Tanna et al., 2017). This is shown in Figure 1.4, where the emergency service is available via an interoperability link between the elderly's smart home system and the emergency service itself.

Third, both IoT and smart cities are *spatially structured*, that is, they have geographical coverage and can be spread along a region, forming alliances, such as the network formed by the smart hospital, the emergency service, and the ambulance.

Regarding the *temporal dimension*, the IoT infrastructure in a smart city may manage many parallel and concurrent events due to the vast number of interactions. In smart cities, real-time data processing is crucial when a system must be notified immediately after a significant event occurs, such as the existing interaction between the smart home environment and the emergency service. In brief, both space and time rule the different types of events taking place in every IoT use case, such as smart cities (Arruda and Bulcão-Neto, 2019).

Moreover, both IoT and smart city systems should provide *intelligent services*, that is, could use AI applications to process the large volume of data that forms a real big data structure. In that sense, big data analytics contributes to making the smart city's constituent systems useful and beneficial to its citizens. In our scenario, traffic sensors spread around the city continuously feed the traffic control system that analyzes such data to guide the ambulance through the safest paths to the hospital.

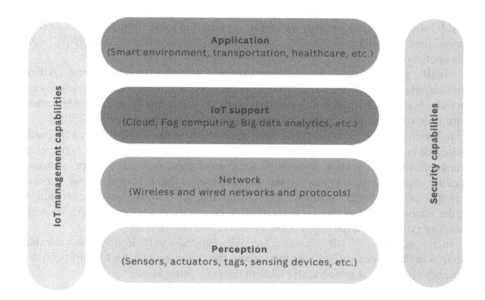

FIGURE 2.4   IoT-oriented architecture for smart cities.

Finally, for both, an *architecture* emerges, that is, a composition of physical objects with their software counterparts that, together with their connections, assume a structure that can be analyzed and monitored in a secure and scalable fashion.

In a smart city, all the systems are networked together to form a real "IoT" directly or indirectly. These things appear in different granularities and dimensions.

Figure 2.4 illustrates a general IoT-based architecture for smart cities, mostly influenced by the ITU-T's reference model for the IoT (ITU-T, 2012).

The *perception* layer of an IoT system is composed of sensors, actuators, tags, and general devices, such as everyday appliances, that are directly or indirectly connected to the network layer.

The *network* layer provides networking and transport capabilities that allow for the secure and low-latency transmission of information, including functions such as authentication, authorization, and accounting, mobility management, and the transport of application-specific data. This layer forms the core of an IoT system, connecting smart devices, network devices, and servers through various networks, including the Internet, sensor networks, and communication networks. While the network layer is crucial to the functioning of an IoT system, it is beyond the scope of this chapter to delve into the details of this layer.

The primary role of the *IoT support* layer is to provide generic functions for diverse applications in a smart city scenario. These functions can include cloud and fog computing services, such as data processing, data storage, networking, and data analytics (Santos et al., 2020). The choice between cloud and fog computing depends on the application's latency requirements, with fog computing being more suitable for low-latency applications (Shrivastava and Pandey, 2020), such as in-home patient health monitoring as previously described. In addition to providing generic support capabilities, the IoT support layer can also cater to the specific requirements of various applications.

Finally, the *application* layer provides the user interface of smart cities' vertical applications with rich visualizations of analysis results. Examples of applications include, but are not limited to, smart government, smart environment, smart transportation, smart utilities, and smart services (e.g., healthcare) (Cui et al., 2018).

IoT management capabilities include generic resources for IoT use cases, such as device management, local network topology management, traffic and congestion management, and resource reservation for critical data. Besides, IoT applications' specific requirements may also be supported. Observe that IoT applications' generic and specific security demands are also considered, for example, privacy protection at the application layer and authorization, authentication, data confidentiality, and integrity protection at the application, network, and device layers.

In the scenario of fall detection for older adults, wearable and non-wearable devices in the home, traffic lights throughout the city, the ambulance's GPS, and the patient's body sensor network all reside in the perception layer.

Smartphones may be adequate for sensing, networking, and processing capabilities. However, the same may not be true for a Doppler radar-based device, particularly in regard to its processing power. The smartphone can collect the elderly person's address, emergency contacts, and movement data (including data from the radar device) and send it to the cloud. The data can be preprocessed, stored, and analyzed through a specific fall detection service in the cloud. If the analysis of the movement data indicates a fall, the fall detection service automatically notifies the smart city emergency service.

However, other designs could also be considered, such as implementing and deploying the fall monitoring service in the elderly person's home. It's important to consider the latency required to transmit the data to the processing center in the cloud and potential issues related to safety and quality of service that may arise if a connection is lost and data is not processed, which could prevent the emergency service from being called and result in a fatal accident. Therefore, architectural alternatives should be explored to ensure that the resulting system meets the requirements and needs of all stakeholders. For example, a fog node serving multiple houses in a city region could acquire, model, and preprocess the georeferenced movement data of older adults and automatically send it to a fall detection service in the cloud.

When the smart city emergency service is notified, it sends an ambulance to the house where the fall occurred. The ambulance is monitored by smart traffic lights throughout the city, which collect and analyze traffic data (such as drivers' speed) as it is collected. To ensure timely response, the traffic lights can make decisions to reorganize routes as needed. The collected data can later be sent to the cloud for more in-depth and long-term analysis, such as for traffic planning purposes.

The smart public health system, which operates in the cloud, collects and analyses information from other healthcare institutions in the city to determine which hospital the emergency team should drive to. The smart traffic lights also assist the ambulance in reaching the hospital.

Finally, the patient's emergency contacts are notified by a private service running within the hospital infrastructure. During the hospitalization period, the patient's vital signs are monitored by a wireless body sensor network that sends the data to a vital sign analyzer service running on the hospital's private cloud. If the analysis of the vital signs indicates a deterioration in the patient's health condition, the service notifies the smartphones of the healthcare professionals in charge.

From the previous discussion and the combination of Figures 1.4 and 2.4, it is clear that IoT and smart cities converge in that the IoT sensing layer serves as the connection point between the smart city system and the public or domestic environment. At the same time, the modeling and reasoning layers are the IoT components that support data representation, service implementation, and interoperability between the systems that make up a smart city. From an architectural perspective, it is important for a smart city system to take into account the different capabilities and manage data accordingly, to ensure effectiveness and efficiency. In the following section, we will discuss some of the challenges to consider for IoT usage in smart cities context.

2.6 CHALLENGES FOR USING IoT AS THE BACKBONE OF SMART CITIES

There are various challenges associated with building IoT-based systems. We address them as follows.

2.6.1 REQUIREMENTS ENGINEERING OF IoT-BASED SYSTEMS

Requirements engineering (RE) is the process of gathering client expectations for a system being developed. A requirement is a description of a client's genuine need that should be satisfied by the resulting system. Requirements elicitation is a key activity in RE where requirements are collected and documented in a requirements specification (RS) document. For example, common requirements for an autonomous car include the ability to drive in automatic pilot, navigate to a destination, search for parking, report road conditions, and provide basic information such as location, heading, speed, and acceleration. These are known as functional requirements (FRs), or actions and behaviors that can be triggered by a user through a mechanism such as a voice command or button. Non-functional requirements (NFRs), also known as "ilities" such as usability, portability, maintainability, and scalability, describe how the system should achieve its duties, and can be measured by metrics such as response time.

RE for IoT-based systems such as smart cities is a complex task due to the new domain, the need for domain-specific knowledge, and the focus on concerns such as fitness of purpose, wireless interoperability, and energy efficiency. NFRs such as security, scalability, and reliability can be of greater concern in certain domains like healthcare or education. Additionally, new NFRs such as context-awareness and mobility are introduced by IoT systems and are not addressed by traditional RE techniques. There are also challenges in integrating and evaluating NFRs in IoT applications. The following subsections provide a panoramic discussion on NFRs considerations for IoT applications.

2.6.1.1   IoT Security

A 2020 survey by Forrester Research (F. Research, 2020) revealed that many businesses in North America struggle to identify, monitor, and secure IoT devices in their operations. Additionally, 67% of surveyed businesses reported experiencing security incidents related to IoT devices. Another study by Forescout Technologies (F. Technologies, 2020) found that smart buildings, medical devices, networking equipment, and VoIP phones are among the riskiest IoT device groups. Six of the top ten highest-risk IoT device types were found to be medical devices and networking equipment. Security is the most frequently discussed NFR for IoT-based systems. Research in IoT security focuses on several key concerns such as:

a. **Sensors vulnerabilities:**
   - Design vulnerability refers to the security weaknesses that occur when proper security measures are not implemented during the development of a device. These vulnerabilities can include a lack of user-friendly interface for changing credentials, control interfaces that do not require user authentication, the use of hard-coded passwords, communication protocols that transmit sensitive information without encryption, and the ability for unauthorized remote firmware updates.
   - Implementation vulnerability refers to the security weaknesses that arise from coding errors that can be exploited during a cyberattack. These vulnerabilities can include buffer overflow, where more data are entered into a buffer than it can handle, and the improper use of random number generators, which can lead to easily guessable security keys.
   - Deployment vulnerability refers to security weaknesses that are introduced by the user during the installation or operation of sensors. These vulnerabilities can include using weak or easily guessed passwords, not changing default passwords, not activating security features, and using counterfeit sensors that may have been tampered with or compromised.

The following software development processes that prioritize security can help address design and implementation vulnerabilities in IoT devices. Examples of such processes include the Open Web Application Security Project (OWASP), Secure Software Development Lifecycle, and Microsoft's Security Development Lifecycle. On the other hand, many practitioners have proposed solutions to the *problem* of default credentials in IoT systems, ranging from the usual recommendation to change credentials, encouraging manufacturers to randomize passwords per device, issuing Manufacturer Usage Description (MUD) specifications allowing manufacturers to specify authorized network traffic, to more advanced and strict ideas such as enacting legislation that regulates the operation of IoT devices (e.g., California legislation [SB-327] that bans default passwords in IoT-connected devices).

b. **Communication channel:** The manner in which IoT devices communicate can vary and may include a range of wireless protocols such as ZigBee, Bluetooth Low Energy (BLE), Wi-Fi, cellular data, and Ethernet. These communication channels are susceptible to malicious interference and disruptions. While using transport encryption protocols such as WEP or WPA2 is advisable, it may not be sufficient. ZigBee and BLE, for example, not only have built-in encryption but also have known vulnerabilities. Therefore, it is recommended to adopt additional standards such as Transport Layer Security (TLS) or Datagram Transport Layer Security (DTLS) to strengthen the communication security whenever possible.

c. **Aggregators:** These are software implementations based on mathematical function(s) that transform groups of raw data received from the IoT devices into intermediate, aggregated data, and then transmit the aggregation result to servers. While an aggregator shall act as an honest but curious entity, whose duty is aggregation and relaying, it may also become a point of attack (e.g., by feeding them fraudulent data or denying them the ability to execute). An adversary may compromise the aggregator to infer the actual data of each connected IoT device, which may compromise the devices' privacy. Existing data aggregation schemes shall be designed with security in mind so that when an adversary forges or modifies a report, the malicious operations should be detected by an aggregator. Aggregators shall also guarantee that the received data are valid and derived from legal entities.

d. **Upgrade process:** IoT devices often require updates to their firmware to add new features or change configurations. These updates should be done through a secure process to ensure that the firmware is coming from a trustworthy source. Machine-to-machine authentication methods can be used by the IoT device to verify the authenticity of the upgrade source before downloading the new firmware image. Cryptographically secure hash validation can also be used to confirm the integrity of the firmware before it is stored on the device. The IoT Firmware Update Architecture (Moran et al., 2019), recently proposed to the Internet Engineering Task Force (IETF), provides guidelines for implementing a secure firmware update process, including specific rules for how device manufacturers should operate.

2.6.1.2 IoT Scalability

Scaling an IoT deployment and infrastructure can be a challenging endeavor. A comprehensive scalability strategy for IoT-based applications shall address the concerns as follows.

a. **Wireless capacity:** Connected devices may generate a deluge of data traffic, which imposes great bandwidth challenges. It is essential to assess whether the wireless system can accommodate a fast-growing number of endpoints that arrive down the line. Metrics such as the number of messages that can be handled per gateway per day can help evaluate the scalability of a wireless network. In addition, with sub-GHz wireless technology, it is possible to segregate IoT networks from other 2.4 GHz legacy systems to mitigate congestion issues.

b. **Network architecture:** The short radio range of many wireless protocols dictates the necessity to distribute the IoT devices and repeaters delicately. Adding or moving nodes can lead to unpredictable performance or troubleshooting challenges. A star topology structure can help in this direction. On the other hand, while the cloud is on the radar to handle massive IoT data streams, a combination with on-premise infrastructure is often called into action to satisfy a balance between cost, performance, security, and scalability. The hybrid workflows and data migration from edge to cloud shall be carefully assessed for scalability in this case. Containerized-based design and microservices make a good fit with hybrid architecture due to their platform-agnostic nature, and the possibility to leverage container orchestration tools such as Kubernetes, to easily deploy, manage, and scale the software to adapt to changing needs.

c. **Data storage:** By embedding sensors into front field environments as well as terminal devices, an IoT network can collect rich sensor data that reflect the real-time environment conditions of the front field and the events/activities that are going on. Since the data is collected in the granularity of elementary event level in a 7×24 mode, the data volume is very high and the data access pattern also differs considerably from traditional business data. This has motivated a new generation of data management solutions, for example, NoSQL database and map-reduce distributed computing framework.

2.6.1.3   Interoperability for IoT

Interoperability is key to unlocking all of the IoT paradigm's potential, including immense technological, economic, and social benefits. Recalling the basic notion, interoperability consists of the ability of two or more systems to exchange data/information and use the data/information exchanged. Then, interoperability is a challenge because interoperability links should enable effective communication between technology elements, and those elements can be highly heterogeneous. The interoperating systems or components maybe been codified in different programming language paradigms, or are deployed in different hardware architectures or operating systems, or the network technologies to communicate are different and not compatible (examples) or the difference can even be at the business level, with a person being represented with a different set of attributes in two different systems. In those cases, *how to establish communication between these systems and enable them to exchange data and use the data exchanged effectively?*

Interoperability is a top challenge possibly preventing IoT from reaching its full potential. The interoperability challenge can manifest itself in several ways: lack of a reference standard, vast heterogeneity of IoT systems, and limited connectivity between different transport protocols, such as Ethernet, Wi-Fi, and ZigBee, are causing an inability to complement and integrate collected data from different IoT devices. Interoperability *per se* is a grand challenge for several disciplines, ranging from computer networks, passing by information systems, and reaching software engineering (Boscarioli et al., 2017; Maciel et al., 2017). For smart cities and IoT, this is not different.

In IoT-based systems (as smart cities), these challenges can be even maximized once: (i) the metrics used to sense the environment can be different among things and demand conversions (e.g., Fahrenheit or Celsius in different devices), (ii) the data representation can be different, which could cause redundancy or inconsistencies in the data being represented in the data modeling step (e.g., ontologies and object-oriented models), and (iii) the reasoning step demands a sort of intelligence to deal with a potential high diversity of types of data and their associated representation (e.g., rule-based and ontology-based reasoning).

Nevertheless, McKinsey Co. estimates that resolving these interoperability issues can unlock more than $4 trillion per year in potential economic impact from IoT use by 2025 (Manyika et al., 2015). It is essential to consider that IoT deployments have specific interoperability needs:

- **Technical:** Ability to use a physical communications infrastructure to transport data.
- **Syntactic:** Ability to share syntax or common information model structures for data and establish a protocol to share the information as specific typed data.
- **Semantic:** Ability to establish data meaning.

Today, the industry is beginning to coalesce around the notion that devices should simply work together in a plug-and-play fashion, and technology standards are progressively becoming popular to foster horizontal interoperability. The Standard for IoT Messaging (MQTT), for example, is an open-source networking protocol that transports messages between devices and has been serving as a *lingua franca* for the wide range of IoT components that can use it to exchange information.

Solutions have been proposed over the years, such as realizing the full interoperability (Maciel et al., 2017) and the automatic synthesis of domain-specific middleware (Costa et al., 2017). In the former, supporting all interoperability levels (syntactic, semantic, pragmatic, dynamic, and organizational or conceptual) for specific domain applications could be considered as full interoperability. In the latter, the idea comprises the automatic synthesis of middleware, that is, a software component capable of abstracting the heterogeneities mentioned above, for specific domains. For instance, in a smart home environment, a software component could analyze the surrounding things and automatically synthesize a component capable of collecting the data representation formats, such as automatically establishing mappings that could support communication without inconsistencies between the involved components. This could avoid inconsistencies in fall detection, for instance. However, both topics are still a matter of investigation, but great promises to solve the problem for IoT and smart cities.

2.6.1.4 IoT Performance

As with any network technology, responsiveness and speed are essential for reliable IoT network operation. Several factors can affect the performance of IoT systems:

- Massive numbers of connected devices to the networks
- Limited bandwidth
- Network topology
- Limited storage, and data utilization capacity
- Malfunctioning devices

Edge computing can help in utilizing the network bandwidth because it forces most bandwidth-hogging processes to run directly on the IoT devices reducing the need to send data back and forth to centralized servers for processing. While traditional WAN links often lack the network intelligence necessary to move IoT data across the network in the most optimal manner, utilizing a software-defined wide area network (SD-WAN) can improve IoT network performance by combining two or more WAN links with AI, allowing data to travel over the optimal path toward its final destination. Network segmentation and adaptive contention window (ACW) are two other performance tactics for IoT-based applications.

2.6.1.5   IoT Usability

The success rate of IoT projects is low, with only 25% succeeding (Cisco Press Release, 2017). Early adopters face challenges with interoperability, while late adopters struggle with the complexity and lack of expertise to set up and navigate through the system. Usability design is particularly challenging in IoT due to the wide range of device types and the need for consistency among user interfaces, easy navigation through security protocols, and limitations such as limited display size and functionality. Additionally, smart home device UIs are often limited to a small set of onboard features, with a broader set of control parameters only accessible remotely via mobile device.

On top of this, users will often invent a use for the device that was not part of your original market concept, such as an IoT-enabled tractor that sends an alert when servicing is required. In this case, a farmer may utilize the feature to also pay employees based on productivity. Establishing the various use cases matters because they identify the usability models specific to the device, task, or user.

2.6.1.6   IoT Discoverability

To realize the vision of truly connected things, there must be mechanisms available for automatic discovery of resources, their properties, and capabilities as well as the means to access them. Device discovery is a complex problem for IoT, but the general problem of discovery within networks has been studied for decades. Bröring et al. presented four categories of IoT discovery technologies (Bröring et al., 2016):

- Discovery of "things," which are in close spatial proximity to a client (<10 cm with NFC and <100 m with BLE).
- Discovery of endpoints of "things" on the network (e.g., mDNS, multicast CoAP, SSDP, and WS-Discovery).
- A central directory is used for the discovery of IoT devices and their resources (e.g., CoRE Resource Directory, XMPP IoT Discovery, HyperCat, Sensor Instance Registry, and SPARQL Endpoint).
- Accessing IoT device metadata once they are discovered (e.g., CoRE Link Format, OGC SOS, and optical markers).

Nevertheless, the discoverability of IoT devices can be contradicting some aspects of security with some illustrating scenarios being discussed in the NIST article on IoT trust concerns (Voas et al., 2018).

2.6.1.7   IoT Mobility

Many IoT devices intrinsically work over mobile systems or evolve toward mobility or both because they move with humans as is the case of smartphones and wearable devices or because they move by themselves as in the case of robots. Based on their locations, these devices are likely to change their IP addresses and networks frequently.

Routing protocols, such as RPL, must reconstruct a tree-like routing topology called the destination-oriented directed acyclic graph (DODAG) each time a node goes off the network or joins the network, which adds to the system a substantial overhead. Fortunately, the research community has been active in developing algorithms to address the attributes of IP mobility management within IPv4 and IPv6.

2.6.1.8   Other Ilities

Deploying IoT systems raise new ethical concerns about personal control. Success in IoT does not only depend on connecting technologies but also addressing new qualities such as humanization or dehumanization in the specific domain of deployment. For example, when building IoT systems for healthcare, it is important to involve all stakeholders in defining the concept of "caring" for the new system. However, IoT technology can also decrease autonomy and give power to corporations focused on financial gain, as seen in the education domain where controlling agents may become the organizations that control the tools used by educators, rather than the educators themselves.

In conclusion, quality requirements have always been a difficult task for the development community, and it becomes even more complex with the introduction of new IoT capabilities and their interactions. We discussed the potential impact of different architectural designs on the quality of a smart home system in this chapter. The complexity of balancing multiple quality attributes in a city-scale system can be even more challenging. Currently, there is a lack of a quality model for smart cities and their IoT-based systems, but this can subject for future research. In particular, attributes such as "protection, security, privacy, and safety" (Spinellis, 2017; IEEE, 2019a), availability (Delécolle et al., 2020), responsiveness (Motta et al., 2019), and others can be further examined to ensure the quality of services provided by smart cities to its users.

Privacy is a key concern in a smart city, as it collects and processes personal, enterprise, and government data constantly, which should not be accessible to unauthorized individuals. In addition to cryptography, anonymization, and access control, two factors are crucial for a comprehensive privacy protection solution: time and space (Arruda and Bulcão-Neto 2019; Perera et al., 2020). For example, in the case of an elderly person who has fallen, emergency services are immediately granted access to their data, but this access is limited to the duration of the hospitalization. Additionally, the closer an IoT device is to the end-user, such as a patient's smartphone, the more challenging it becomes to protect their privacy before sensitive data "leaves" their home (Cui et al., 2018).

2.6.2   A Standard Reference Architecture for Smart Cities

The IEEE's IoT architecture working group recently developed an architectural framework for the IoT called P2413, motivated by concerns commonly shared by systems stakeholders across multiple IoT domains (IEEE, 2019a). In general, these concerns relate to disjointed, domain-specific, and often redundant standardization efforts. The P2413 incorporates a collection of architecture viewpoints elaborated to form the framework description body: a reference model, a reference architecture, and a blueprint for data abstraction.

An active project of the same IEEE working group leveraged on the IEEE P2413 is the standardization of a Reference Architecture for Smart Cities (RASC) (IEEE, 2019b). RASC's goals include defining different vertical applications in the smart city (e.g., water and waste management, environment monitoring, smart buildings, and eHealth), commonalities between these vertical applications, and an intelligent and integrated operations center in a smart city use case. Besides, RASC also proposes a four-tiered architecture for smart cities (device, communication network, IoT platform, and application) and defines the relationships with attributes specific to cloud computing technologies and big data analysis. By this writing, RASC is still an IEEE draft standard, thus under regular revisions.

2.6.3 SOFTWARE-ENGINEERING SMART CITIES

Networking infrastructure is a necessary investment to make a smart city a reality, but it is not sufficient on its own (Weyrich and Ebert, 2016). Software plays a crucial role in smart cities, covering various areas such as data acquisition, communication, representation, storage, reasoning, distribution, retrieval, and visualization. The multidisciplinary nature of IoT and the numerous stakeholders in a smart city pose challenges to current software engineering practices (Santana et al., 2018; Motta et al., 2019).

In a smart city use case, requirements should be gathered from citizens and system stakeholders using a more comprehensive approach that includes business, usage, functional, and implementation viewpoints, as Fraile and colleagues propose for Digital Manufacturing Platforms (Fraile et al., 2019). For example, the business viewpoint focuses on a value-driven model with quantitative performance indicators such as business value, expected return on investments, and maintenance costs. The usage perspective identifies the parties, actors, roles, tasks, and workflows involved in different smart city use cases. The functional viewpoint involves decomposing each constituent system into functional parts and describing its structure, interrelations, interfaces, and interactions between internal components and other systems. It should also include data model definitions, security requirements, and alignment of tasks to functional and implementation components. The implementation viewpoint should provide details on how an asset will be deployed to the computing infrastructure. Coordinating all these multiple points of view is crucial in a smart city development project, however, representing, describing, and integrating systems in light of software requirements is still an ongoing challenge (Motta et al., 2019).

As Spinellis (2017) pointed out, designing and constructing integrated IoT systems such as smart cities requires complex adaptation layers due to the dynamic nature of requirements. The scale of a smart city also poses challenges to software verification and validation techniques, including simulation (Larrucea et al., 2017; Graciano Neto et al. 2018). With the increasing adoption of IoT technologies and real-world smart cities experiences, more research on software engineering is needed, including general models, methodologies and tool support (Morin et al., 2017; Zambonelli et al., 2017; Santana et al., 2018; Motta et al., 2019).

2.7 FINAL REMARKS

In this chapter, we explored the connection between the IoT and smart cities. We discussed the overall structure of a smart city and how IoT applications are integrated, highlighting their close relationship. We used the example of a smart home system for detecting falls in elderly individuals and linking them to emergency services to illustrate how IoT sensing is the link between smart cities and the public/domestic environment. We also highlighted the ongoing research opportunities in this field.

Smart cities and IoT are closely related, with the latter being the backbone of the former. Both share similar characteristics and challenges. As IoT technologies and their challenges, such as full interoperability, are addressed, smart cities will greatly benefit and be able to offer more advanced services to society. The next chapter discusses systems of systems, the mother class of smart cities.

REFERENCES

Alampalli, S., & Pardo, T. (2014). A study of complex systems developed through public private partnerships. In Proceedings of the 8th International Conference on Theory and Practice of Electronic Governance – ICEGOV '14, 442–445. https://doi.org/10.1145/2691195.2691212

Anishchenko, L. N., Zhuravlev, A. V., Razevig, V. V., & Chizh, M. A. (2019, June). Low-cost portable bioradar system for fall detection. In 2019 PhotonIcs & Electromagnetics Research Symposium-Spring (PIERS-Spring) (pp. 3566–3570). IEEE. Rome, Italy.

Arruda, M. F., & Bulcão-Neto, R. F. (2019). Toward a lightweight ontology for privacy protection in IoT. In Proceedings of the 34th ACM/SIGAPP Symposium on Applied Computing (SAC '19). Association for Computing Machinery (pp. 880–888). New York, NY, USA. https://doi.org/10.1145/3297280.3297367

Asiag, J. J. (2020). 8 Smart Cities Lead the Way in Advanced Intelligent Transportation Systems. Otonomo. https://otonomo.io/blog/smart-cities-intelligent-transportation-systems/

Boscarioli, C., Araujo, R. M., & Maciel, R. S. P. (2017). I GranDSI-BR – Grand Research Challenges in Information Systems in Brazil 2016 – 2026. Special Committee on Information Systems (CE-SI). Brazilian Computer Society (SBC). 184p ISBN: [978-85-7669-384-0]

Bröring, A., Datta, S. K., & Bonnet, C. (2016). A categorization of discovery technologies for the Internet of Things. In Proceedings of the 6th International Conference on the Internet of Things (pp. 131–139). Stuttgart, Germany.

Bulcão-Neto, R., Teixeira, P., Lebtag, B., Graciano-Neto, V., Macedo, A., & Zeigler, B. (2023). Simulation of IoT-oriented fall detection systems architectures for in-home patients. IEEE Latin America Transactions, 21(1), 16–26. https://latamt.ieeer9.org/index.php/transactions/article/view/6863

Burns, R., Stevens, J. A., & Lee, R. (2016). The direct costs of fatal and non-fatal falls among older adults – United States. Journal of Safety Research, 58, 99–103.

Carvalho, R. M., Andrade, R. M., & De Oliveira, K. M. (2018, May). Towards a catalog of conflicts for HCI quality characteristics in UbiComp and IoT applications: Process and first results. In 2018 12th International Conference on Research Challenges in Information Science (RCIS) (pp. 1–6). IEEE, Brighton, United Kingdom.

Cavalcante, E., Cacho, N., Lopes, F., & Batista, T. (2017). Challenges to the development of smart city systems: A system-of-systems view. In Proceedings of the Brazilian Symposium on Software Engineering, 244–249.

Chaudhuri, S., Thompson, H., & Demiris, G. (2014). Fall detection devices and their use with older adults: A systematic review. Journal of Geriatric Physical Therapy, 920 (37), 178.

Cisco Press Release. (2017). Cisco Survey Reveals Close to Three-Fourths of IoT Projects Are Failing. https://newsroom.cisco.com/c/r/newsroom/en/us/a/y2017/m05/cisco-survey-reveals-close-to-three-fourths-of-iot-projects-are-failing.html. Last Accessed, May 2022.

Costa, F. M., Morris, K. A., Kon, F., & Clarke, P. J. (2017). Model-driven domain-specific middleware. In Proceedings of the International Conference on Distributed Computing Systems 2017 (pp. 1961–1971). Atlanta, GA.

Cui, L., Xie, G., Qu, Y., Gao, L., & Yang, Y. (2018). Security and privacy in smart cities: Challenges and opportunities. IEEE Access, 6, 46134–46145. https://doi.org/10.1109/ACCESS.2018.2853985

DeFranco, J., & Kassab, M. (2021). What Every Engineer Should Know About the Internet of Things. CRC Press.

Delécolle, A., Lima, R. S., Graciano Neto, V. V., & Buisson, J. (2020). Architectural strategy to enhance the availability quality attribute in system-of-systems architectures: A case study. In Proceedings of the System of Systems Engineering Conference 2020 (pp. 93–98). Budapest, Hungary.

F. Research. (2020). State of enterprise IoT security in North America: Unmanaged and unsecured. https://www.armis.com/wp-content/uploads/2020/03/Armis-State-of-Enterprise-IoT-Security-INFO.pdf

F. Technologies. (2020). The enterprise of things security report: The state of IoT security. https://www.forescout.com/the-enterprise-of-things-security-report-state-of-iot-security/#:~:text=The%20State%20of%20IoT%20Security,-In%20this%20first&text=Some%20of%20the%20key%20findings,medical%20devices%20and%20networking%20equipment.

Fraile, F., Sanchis, R., Poler, R., & Ortiz, A. (2019). Reference models for digital manufacturing platforms. *Applied Sciences*, 9(20), 4433.

Giannoutakis, K. M., Spanopoulos-Karalexidis, M., Filelis Papadopoulos, C. K., & Tzovaras D. (2020). Next generation cloud architectures. In: Lynn T., Mooney J., Lee B., Endo P. (eds) The Cloud-to-Thing Continuum. Palgrave Studies in Digital Business & Enabling Technologies. Palgrave Macmillan, Cham. https://doi.org/10.1007/978-3-030-41110-7_2

Graciano Neto, V. V., Manzano, W., Kassab, M., & Nakagawa, E. Y. (2018). Model-based engineering & simulation of software-intensive systems-of-systems: Experience report and lessons learned. In Proceedings of the European Conference on Software Architecture (Companion) (27, pp. 1–27:7). Madrid, Spain.

Graciano Neto, V. V., Santos, R. P., Viana, D., & Araujo, R. (2020). Towards a conceptual model to understand software ecosystems emerging from systems-of-information systems. In: Santos R., Maciel C., Viterbo J. (eds) Software Ecosystems, Sustainability and Human Values in the Social Web. WAIHCWS 2017, WAIHCWS 2018. Communications in Computer and Information Science, vol 1081. Springer, Cham. https://doi.org/10.1007/978-3-030-46130-0_1

Gutierrez-Madroñal, L., Blunda, L. L., Wagner, M., & Medina-Bulo, I. (2019). Test event generation for a fall-detection IoT system. IEEE Internet of Things Journal, 6, 6642–6651.

Harrison, C., Eckman, B., Hamilton, R., Hartswick, P., Kalagnanam, J., Paraszczak, J., & Williams, P. (2010). Foundations for smarter cities. IBM Journal of Research and Development, 54(4), 1–16.

Hung, M. (2017). Leading the IoT. Gartner insights on how to lead in a connected world. Gartner. https://www.gartner.com/imagesrv/books/iot/iotEbook_digital.pdf

IEEE P2413. (2019a). Standard for an Architectural Framework for the Internet of Things (IoT). IEEE Standards Association, 21 May. Accessed November 2020. https://standards.ieee.org/standard/2413-2019.html

IEEE P2413.1. (2019b). Standard for a reference architecture for smart City (RASC). IEEE Standard Association. Accessed November 2020. https://standards.ieee.org/project/2413_1.html

ISO. (2014). ISO 37120 – Sustainable development of communities – Indicators for city services and quality of life.

ITU-T. (2012). Recommendation Y.2060: Overview of the Internet of things. http://handle.itu.int/11.1002/1000/11559

ITU-T Focus Group on Smart Sustainable Cities. (2014). Smart Sustainable Cities: An Analysis of Definitions. Focus Group Technical Report, Geneva, Switzerland, Tech. Rep. FG-SSC-10/2014.

Kassab, M., DeFranco, J., & Laplante, P. (2020). A systematic literature review on Internet of Things in education: Benefits and challenges. Journal of Computer Assisted Learning, 36(2), 115–127.

Khanna, A., & Kaur, S. (2020). Internet of Things (IoT), applications and challenges: A comprehensive review. Wireless Personal Communications, 114, 1687–1762. https://doi.org/10.1007/s11277-020-07446-4

Laplante, P. A., Kassab, M., Laplante, N. L., & Voas, J. M. (2017). Building caring healthcare systems in the Internet of Things. IEEE Systems Journal, 12(3), 3030–3037.

Larrucea, X., Combelles, A., Favaro, J., & Taneja, K. (2017). Software engineering for the internet of things. IEEE Software, 34(1), 24–28.

Lear, E., Droms, R., & Romascanu, D. (2019). Manufacturer usage description specification (no. rfc8520). Tech. Rep.

Lynn, T., Endo, P. T., Ribeiro, A. M. N. C., Barbosa, G. B. N., Rosati, P. (2020). The Internet of Things: Definitions, key concepts, and reference architectures. In: Lynn T., Mooney J., Lee B., Endo P. (eds) The Cloud-to-Thing Continuum. Palgrave Studies in Digital Business & Enabling Technologies. Palgrave Macmillan, Cham. https://doi.org/10.1007/978-3-030-41110-7_1

Maciel, R. S. P., David, J. M. N., Claro, D. B., & Braga, R. (2017). Full Interoperability: Challenges and Opportunities for Future Information Systems. Grand Challenges in Information Systems for the next 10 years (2016-2026), SBC, 107–118.

Mahalank, S. N., Malagund, K. B., & Banakar, R. M. (2016). Non Functional Requirement Analysis in IoT based smart traffic management system. In 2016 International Conference on Computing Communication Control and Automation (ICCUBEA) (pp. 1–6). IEEE, Pune, India.

Manyika, J., Chui, M., Bisson, P., Woetzel, J., Dobbs, R., Bughin, J., & Aharon, D. (2015). Unlocking the potential of the Internet of Things. McKinsey Global Institute, 1.

Manzano, W., Graciano Neto, V. V., & Nakagawa, E. Y. (2020). Dynamic-SoS: An approach for the simulation of systems-of-systems dynamic architectures. The Computer Journal, 63(5), 709–731.

Maranhão, G. M., & Bulcão-Neto, R. (2016). A semantic filtering mechanism geared towards context dissemination in ubiquitous environments. Journal of Universal Computer Science, 22, 1123–1147.

MarketsandMarkets. (2022). IoT in Manufacturing Market worth $87.9 billion by 2026. https://www.marketsandmarkets.com/PressReleases/iot-manufacturing.asp

Mauldin, T. R., Canby, M. E., Metsis, V., Ngu, A. H., & Rivera, C. C. (2018). Smartfall: A smartwatch-based fall detection system using deep learning. Sensors, 18, 3363.

Mohanty, S. P., Choppali, U., & Kougianos, E. (2016). Everything you wanted to know about smart cities: The Internet of Things is the backbone. IEEE Consumer Electronics Magazine, 5(3), 60–70.

Moran, B., Meriac, M., Tschofenig, H., & Brown, D. (2019). A Firmware Update Architecture for Internet of Things Devices. Internet Engineering Task Force, Internet-Draft.

Morin, B., Harrand, N., & Fleurey, F. (2017). Model-based software engineering to tame the IoT jungle. IEEE Software, 34(1), 30–36. https://doi.org/10.1109/MS.2017.11

Motta, R., de Oliveira, K., & Travassos, G. (2019). On challenges in engineering IoT software systems. Journal of Software Engineering Research and Development, 7, 5:1–5:20. https://doi.org/10.5753/jserd.2019.15

Mulley, G. (2001). Falls in older people. Journal of the Royal Society of Medicine, 94(4), 202.

Perera, C., Barhamgi, M., Bandara, A. K., Ajmal, M., Price, B., & Nuseibeh, B. (2020). Designing privacy-aware Internet of Things applications. Information Sciences, 512, 238–257, ISSN 0020-0255. https://doi.org/10.1016/j.ins.2019.09.061

Perera, C., Barhamgi, M., De, S., Baarslag, T., Vecchio, M., & Choo, K. R. (2018). Designing the sensing as a service ecosystem for the Internet of Things. IEEE Internet of Things Magazine, 1(2), 18–23. https://doi.org/10.1109/IOTM.2019.1800023

Perera, C., Zaslavsky, A., Christen, P., & Georgakopoulos, D. (2014). Context aware computing for the Internet of Things: A survey. IEEE Communications Surveys & Tutorials, 16(1), 414–454, First Quarter 2014. https://doi.org/10.1109/SURV.2013.042313.00197

Ruiz-López, T., Noguera, M., Rodríguez Fórtiz, M. J., & Garrido, J. L. (2013). Requirements systematization through pattern application in ubiquitous systems. In Ambient Intelligence-Software and Applications (pp. 17–24). Springer, Heidelberg.

Santana, E. F. Z., Chaves, A. P., Gerosa, M. A., Kon, F., & Milojicic, D. S. (2018). Software platforms for smart cities: Concepts, requirements, challenges, and a unified reference architecture. ACM Computing Surveys, 50(6), Article 78, 37 pages. https://doi.org/10.1145/3124391

Santos G. L., Monteiro K. H. C., & Endo P. T. (2020). Living at the edge? Optimizing availability in IoT. In: Lynn T., Mooney J., Lee B., Endo P. (eds) The Cloud-to-Thing Continuum. Palgrave Studies in Digital Business & Enabling Technologies. Palgrave Macmillan, Cham. https://doi.org/10.1007/978-3-030-41110-7_5

Sheng, X., Tang, J., Xiao, X., & Xue, G. (2013). Sensing as a service: Challenges, solutions and future directions. IEEE Sensors Journal, 13(10), 3733–3741. https://doi.org/10.1109/JSEN.2013.2262677

Shrivastava, R., & Pandey, M. (2020). Real time fall detection in fog computing scenario. Cluster Computing, 23, 2861–2870. https://doi.org/10.1007/s10586-020-03051-z

Spinellis, D. (2017). Software-engineering the Internet of Things. IEEE Software. 34(1), 4–6. https://doi.org/10.1109/MS.2017.15

Stevens, J. A., Mahoney, J. E., & Ehrenreich, H. (2014). Circumstances and outcomes of falls among high risk community-dwelling older adults. Injury Epidemiology, 1(1), 5. https://doi.org/10.1186/2197-1714-1-5

Tanna, R. D., Kumar, K. S. V., & Karthika, S. (2017). Analytics as a service for beginners. In Proceedings of the 2017 International Conference on Computational Intelligence in Data Science (ICCIDS) (pp. 1–6). Chennai, India. https://doi.org/10.1109/ICCIDS.2017.8272659

Teixeira, P. G., Lebtag, B. G. A., Dos Santos, R. P., Fernandes, J., Mohsin, A., Kassab, M., & Neto, V. V. G. (2020). Constituent system design: A software architecture approach. In 2020 IEEE International Conference on Software Architecture Companion (ICSA-C) (pp. 218–225). IEEE, Salvador, Brazil.

Voas, J., Kuhn, R., Laplante, P., & Applebaum, S. (2018) "Internet of Things (IoT) trust concerns." NIST Cybersecurity White Paper, 1, 1–50, Oct. 17,. https://csrc.nist.gov/CSRC/media/Publications/white-paper/2018/10/17/iot-trust-concerns/draft/documents/iot-trust-concerns-draft.pdf

Weyrich, M., Ebert, C. (2016). "Reference architectures for the Internet of Things," IEEE Software, 33(1), 112–116. doi:10.1109/MS.2016.20.

Zambonelli, F. (2017). Key abstractions for IoT-oriented software engineering. IEEE Software, 34(1), 38–45.

3 Smart Cities as Complex Systems
A Systems-of-Systems Approach*

3.1 INTRODUCTION

During the 1990s, the popular MacGyver TV show featured a protagonist who possessed a unique skill set – the ability to create complex objects such as guns or bombs by combining simple items such as bubble gum, powder, and cables. While exaggerated, this type of ingenuity is precisely what's needed to build smart cities. To achieve our goals, such as efficient traffic flow or rapid emergency response, we must acquire and/or integrate systems that possess the necessary functionalities to deliver complex services. The resulting behavior will be the sum of individual system functionalities. For instance, in a smart traffic control system (STCS), autonomous cars can leverage their respective sensing and communication capabilities to interact and maintain a steady flow of traffic, reducing the likelihood of congestion.

Smart cities are an example of a class of systems with unique characteristics and sometimes referred to as systems of systems (SoS). Smart cities can involve software systems for the management of diverse elements, including traffic, transportation, water/gas supply, power generation and supply, health services, waste management, and retails; besides, cars, people, companies, organizations, and infrastructures (e.g., roads, telecommunication, wireless network, transportation, buildings, bridges, tunnels, parks/gardens, canals, and rivers) are also involved. All these elements can be characterized as *constituents* (or constituent systems) of an SoS, as they are independent operationally and managerially and have their own missions and goals, while they are required to connect among them to provide larger and unique services for smart cities. From this perspective, smart cities can be considered as good examples of SoS.

Designing smart cities demands a critical skill – the capability to amalgamate existing systems and their functionalities to create more sophisticated systems and behaviors that surpass what individual systems can offer. Essentially, it involves building an SoS that can deliver complex services that individual systems cannot provide on their own.

Interestingly, inner systems of a smart city are also classified as smaller SoS. For example, a smart grid is an example of an SoS.[1] In a smart grid, renewable resources

* Portions of this chapter were contributed by Wallace Manzano (0000-0001-5602-3023), Rodrigo Pereira dos Santos (0000-0003-4749-2551), and Elisa Yumi Nakagawa (0000-0002-7754-4298).

DOI: 10.1201/9781003348542-3

such as photovoltaic cells or wind turbines are controlled by software and can be, each one of them, a constituent system and to become part of an SoS. Each of these systems might generate energy, respectively, from irradiation or wind velocity, while battery storage (also controlled by software and another constituent system) can preserve the energy for later use. The interplay among these constituents might lead to more elaborated behaviors as a result, such as a (i) *self-regulated balance between energy production and consumption*, (ii) *energy use optimization* and (iii) *grid stability* (Silva et al., 2022). However, this combination of functionalities, in the form that is done nowadays, might lead to waste of resources and non-optimized architectures impacting on budget and results delivered.

The concept of SoS has a long history dating back to the early days of systems engineering, with early work on the topic arising in the 1960s and 1970s. The U.S. Department of Defense (DoD) was one of the responsible entities for the popularization of the term in the early 2000s. The DoD was facing a growing challenge in managing the integration and coordination of complex systems, such as weapons systems and command and control networks, that were made up of many independent systems, or "constituents." The term "SoS" was used to describe these complex systems that resulted from the interaction of multiple independent systems.

Pragmatically speaking, SoS are alliances of multiple independent systems that work together to achieve some major goal. SoS are not just a group of separate systems, but they are also a result of companies competing (or cooperating) globally (Fernandes et al., 2020). This competition has led companies to work together in partnerships to meet business needs and take advantage of opportunities (something frequently called as *coopetition*) (Graciano Neto et al., 2019b; Basso et al., 2022). For example, a big company like Amazon might partner with smaller companies to offer a wider range of products. These partnerships happen not only at the business level but also at the technical level. This means that the systems of different companies are connected to each other to enable new behaviors, such as allowing other companies to sell their products through Amazon's platform. This type of collaboration requires different groups to work together and share information, with different goals and objectives in order to achieve things that a single system could not do on its own.

Over the past two decades, SoS has become a popular topic of research in the fields of systems engineering, computer science, and management science. The concept has been applied in various domains, such as aerospace, transportation, healthcare, and emergency management. However, the concept of SoS has also gained popularity in the civil domain, in smart cities.

SoS has been implemented in some smart cities around the world, such as Singapore (Payne et al., 2020) and Barcelona (Gascó-Hernandez, 2018), to improve various aspects of city life, such as transportation, energy, and public safety. For instance, Singapore has implemented an intelligent transport system (ITS) that utilizes data from various sources, such as traffic cameras, sensors, and global positioning system (GPS)-enabled vehicles, to optimize traffic flow and reduce congestion (Axelsson and Nylander, 2018). In Barcelona, the city has implemented a smart grid system that allows for real-time monitoring and management of energy consumption, as well as the integration of renewable energy sources. These examples illustrate the potential of SoS in smart cities to create more efficient, effective, and sustainable urban environments.

The main goal of a chapter on SoS is to characterize and define the concept of SoS, including its unique characteristics and types, provide examples of SoS, outline means to design, develop, and evolve SoS, address challenges and potential difficulties in engineering SoS, and provide conclusions and future work.

3.2 SYSTEM OF SYSTEMS: CONCEPT, CHARACTERISTICS, MISSION, EMERGING BEHAVIOR, AND QUALITIES

3.2.1 THE CONCEPT OF SYSTEM AND ITS RELATION TO SoS

SoS, as a concept, was born in the systems engineering domain, an interdisciplinary branch of engineering that enables the realization of successful systems (INCOSE, 2012). Systems engineering, whose term was coined in the 1940s, focuses on design and management of complex engineering systems, including spacecraft, aircraft, automobile, and robots, only to mention a few of them.

Systems engineering understands a system as a combination of interacting elements organized to achieve one or more stated purposes, while a system architecture is the fundamental conception of a system in its environment embodied in elements, its relationships to each other and to the environment, and principles guiding system design and evolution (Dickerson and Mavris, 2011), as mentioned in Section 3.2. In short, a system is understood under a holistic viewpoint, that is, as a whole entity that comprises a myriad of elements from a diverse nature, involving humans, hardware, software, and possibly other components, as weapons, meteorological stations, satellites, GPS, embedded computers, and hardware–software associations.

Systems have evolved from single systems to SoS thanks to the improvements in the communication and computational capacities, resulting in the area of SoS engineering (SoSE). In parallel, software engineering also evolved and emerged as a discipline focused specifically on the development and maintenance of software systems, together with some specificities, such as for embedded, cyber-physical, and real-time systems. Current and next generation of SoS have become more and more tightly driven by software and, therefore, require approaches that better fit to their planning, design, development, and evolution. Knowledge and know-how from software engineering have been leveraged to address the unique characteristics of software-intensive SoS. Hence, software-intensive SoSE refers to the study and application of engineering, that is, a systematic, disciplined, and quantifiable approach to the development, operation, and evolution of SoS where software contributes essential influences and features to such systems as a whole.

As mentioned in Section 3.1, a SoS is an alliance of multiple systems working together to accomplish a set of common goals. As such, it is necessary to understand what can be considered a system. The general systems theory (GST) states that a *system* is made up of different parts that work together to achieve a certain goal, using resources and producing results. This idea of a system can be applied to understand complex structures in many different fields, such as biology, social science, and engineering. Examples of systems include following:

- In the field of biology, the human body is an example of a system. It is made up of different parts, such as the heart, lungs, and muscles, that work together to keep the body alive and functioning. It also interacts with its environment, such as taking in food and oxygen and expelling waste.
- In the field of vehicles engineering, a car is an example of a system. It has different parts, such as the engine, transmission, and wheels, that work together to make the car move. It also has systems inside of it, such as the electrical system and fuel system, that help it operate.
- In the field of social science, a family is an example of a system. It is made up of different members who interact with one another to meet the needs of the family. It also has rules and norms in place to maintain order and stability.
- In the field of mechanical engineering, a jet engine is an example of a system. It is made up of different parts, such as compressors, turbines, and combustors, that work together to generate thrust and power to the aircraft. It also has systems inside of it, such as the fuel and lubrication systems, that help it operate.
- In the field of electrical engineering, a power grid is an example of a system. It is made up of different power plants, substations, and transmission lines that work together to generate and distribute electricity to the customers. It also has control systems in place to regulate the flow of power and maintain stability on the grid.
- In the field of chemical engineering, a water treatment plant is an example of a system. It is made up of different processes, such as coagulation, sedimentation, filtration, and disinfection, that work together to purify water for safe consumption. It also has monitoring and control systems in place to ensure the water meets safety standards.
- In the field of civil engineering, a building is an example of a system. It is made up of different systems, such as structural, HVAC, plumbing, and electrical systems, that work together to create a safe and comfortable environment for the occupants. It also has monitoring and control systems in place to ensure the building is operating efficiently and safely.
- In the field of aerospace engineering, a satellite is an example of a system. It is made up of different subsystems, such as power, propulsion, communication, and payload, that work together to perform specific functions in space. It also has control systems in place to maintain its orientation and trajectory.

For the purpose of this book, we can consider, as a system, any complex product that derives from an engineering project, such as an airplane, a vehicle, an energy power distribution system, a mechanical mechanism, a mobile phone, or even a watch. Here, particularly, we are interested in a specific type of system: the *software-intensive systems*, that is, those systems that have software and hardware elements to operate, whose software crosscuts their entire lifecycle (ISO 42010). Examples of software-intensive systems include any engineered system that has software and hardware elements, being able to connect with others using a communication technology, such as a smart watch, a smartphone, a smart TV, an autonomous car, a smart traffic system, and so on. Observe that the presence of software elements in an engineered system it is what makes it *smarter*, refining the precision of their operation. Even brake systems have used software to improve precision in their actions.

SoS are exactly the evolution of the system concept. In the context of " SoS,"[2] a "constituent system" refers to any individual system that is (or can be) a part of the larger SoS. A SoS is made up of multiple individual systems that work together to achieve a common goal or perform a specific function. These individual systems can be composed of hardware, software, and/or human elements, and they may be developed by different organizations or entities. They can be autonomous systems that can function independently, but when combined, they form a larger and more complex system that offers new capabilities and functionalities. Constituent systems in a SoS can be connected via physical interfaces, networks, or through a common operating environment. It is worth noting that an SoS can itself be comprised entirely of another SoS, as illustrated in Figure 3.1. For instance, smart traffic systems, smart grids, smart houses, and smart buildings are individual SoS, each composed of several other systems. Yet, these SoS are also integral components of a larger SoS – the smart city. Thus, an SoS can be structured across multiple levels, comprising individual systems or smaller SoS as constituents.

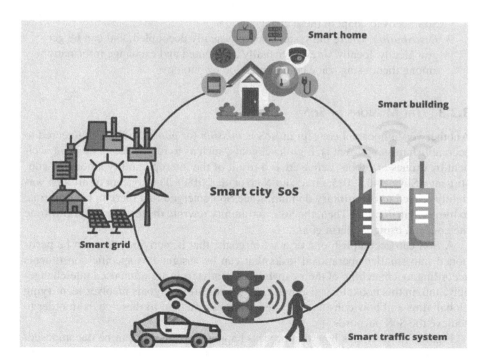

FIGURE 3.1   A smart city SoS formed by constituents that are, themselves, other SoS.

3.2.2 CHARACTERISTICS OF SoS

Besides the characteristics of being composed of other systems or SoS, Maier (1998) defined SoS as a set of five characteristics: emergent behavior, evolutionary development, distribution, and operational and managerial independences of its constituent systems. These characteristics, which were originally defined from a system engineering perspective, are still widely accepted as defining SoS today. Two of these characteristics relate to the nature of the constituent systems:

1. *Operational independence*, where each system contributes to the SoS but also operates independently and preserves its own missions and resources
2. *Managerial independence*, where different constituent systems in a same SoS can be owned by different entities, people or companies, and bring with them their own concerns, restrictions, and potentially conflicting business rules and intents

Additionally, three characteristics relate to the relationships among the constituent systems:

1. *Emergent behavior*, where high-level functionalities and services are offered by the entire SoS because of interactions and cooperation among the systems
2. *Evolutionary development*, where the systems continually evolve and the SoS needs to adapt to these changes
3. *Distribution*, where the systems are physically decoupled, and can be geographically, locally, or even virtually distributed and exchange information among them using some communication technology

3.2.3 THE MISSIONS OF SoS

Another very important concept in SoS is *mission* (or *goal*). SoS are engineered to accomplish missions, that is, high-level goals, such as to rescue someone in an accident by drones or robots, achieved as a result of the interoperability among its constituents (Silva et al., 2015; Graciano Neto et al., 2018c). The concept of mission was tightly related to the military domain, since SoS emerged first there to later migrate to the civilian domain. Then, the SoS community rewrote the term to better illustrate the concept, using the term *goal*.

A SoS can accomplish one or a set of goals, that is, activities that can be partitioned into smaller operational tasks that can be assigned to specific constituents according to a matching of their capabilities. Both terms are often used interchangeably, and, in this book, they also will. The specification of goals involves identifying global aims and how constituent systems can be assigned to these goals in order to achieve the SoS purposes.

Figure 3.2 illustrates how this process happens. A mission can be decomposed/refined in sub-missions. Those sub-missions can then also be refined until the granularity of what is known as a task, that is, a concept close to a requirement, which

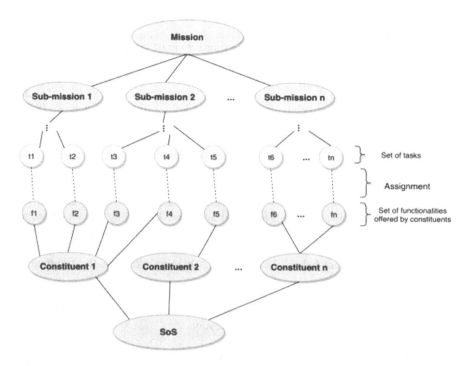

FIGURE 3.2 Mission/goal refinement process.

in turn consists of a need that can be resolved by/implemented as a functionality available in one of more of the constituents (the concept of requirements was further explored in Section 2.6.1).

Analogously, the SoS also splits into constituents and their respective functionalities so that they reach a granularity similar to that of the refined tasks and the engineer can do one or more assignments, that is, to assign constituents functionalities to one or more tasks in a way that all the tasks can be accomplished and the mission itself is also accomplished. The goal can be successively refined into smaller goals until reaching a granularity level that matches the capabilities/functionalities offered by the constituent systems of a SoS.

Examples of missions that can be accomplished by a SoS include "rescue a person." That mission can be refined in other smaller steps, such as (i) find the person, (ii) evaluate if he/she has injuries, (iii) evaluate if it is safe to remove the person from the local perspective, (iv) evaluate the vital signs and transmit them to the command and control (C2) agency, and (v) remove the person from the risky local. Each of these steps can be further refined into even smaller steps or assigned to different entities, such as rescuers, swimmers, or even rescue robots. Different systems with their own functionalities can be involved to accomplish the mission, such as an aquatic mobile sonar to accomplish step (i), a medical autonomous system to accomplish tasks (ii) and (iv), the rescue chief team through a drone with camera to accomplish task (iii), and a robot can perform actions to accomplish task (v).

3.2.4 EMERGENT BEHAVIORS

Emergent behaviors are particularly interesting for smart cities context, since they materialize the services being provided by the constituents to the citizens. Some initiatives have even employed artificial intelligence (AI) to predict the behaviors that could be provided by a SoS, once the number of constituents can be so high that the realization of the emergent behaviors can be complex for humans to achieve (Silva et al., 2022); this is exactly what characterizes a complex system, that is, a collective system with many independent and interdependent parts that interact in a non-linear manner (behavior cannot be expressed as a simple sum of the activity of individual parts) and have interdependencies that are difficult to describe, predict, and design (Alampalli and Pardo, 2014).

An emergent behavior can be classified under two perspectives (Mittal and Rainey, 2015, Graciano Neto et al., 2017a): *intention* and *type*. Regarding the intention, an emergent behavior can be *predicted* or *unpredicted* (Chalmers, 2006), as follows:

- *Predicted* emergent behavior consists of behaviors intentionally designed to emerge at runtime,
- *Unpredicted* emergent behavior corresponds to that one that emerges as a collateral effect of specific conditions or runtime configurations, with the potential to cause losses to the SoS operation.

Considering the type, four categories exist (Mittal and Rainey, 2015): *Simple*, *predicted*, *strong*, and *spooky,* explained as follows:

1. *Simple* emergent behaviors are emergent properties readily intended for the SoS. They are produced in lower complexity through models that abstract the SoS (only intentional behaviors emerge since the model is overly simple).
2. *Predicted* emergent behaviors are those readily and consistently reproducible in simulations of the system, but not in static models. They are partially predicted in advance (desired behaviors are predicted, but undesired can also appear).
3. *Strong* emergent behaviors are consistent with SoS known properties, but are not reproducible in any model of the system. Direct simulations may reproduce the behavior, but inconsistently, and simulations do not predict where the property will occur (desired behaviors exist, but unpredicted behaviors are the majority).
4. Finally, *spooky* emergent behaviors, which are inconsistent with known properties of the SoS, are not reproducible or subject to simulation (a natural emergence, such as life itself, not predicted).

3.2.5 SOS SOFTWARE ARCHITECTURES AND QUALITY ATTRIBUTES

Besides the essential characteristics highlighted by different authors, some concepts are also determinant to better understand SoS and, consequently, smart cities, such as software architecture and quality attributes (in particular, interoperability, as already covered in the gray box of Section 1.2 in Chapter 1). The notion of architecture sometimes brings confusion to other engineers than civil engineers. Part of the common understanding of architecture has much of art and design, since it can regard visual forms we can observe in the urban environments. However, in a more deep perspective, the notion of architecture is related to two main concepts: *structure* and *behavior*. Structure is related to the composition of the parts that form the whole, considering the inner parts and how they are linked to each other. For instance, the structure of a building regards its main parts that compose the format being displayed and the elements that "glue" all those parts; the form itself is a property that emerges as a result of the parts being united. Behavior regards the function it serves as a result of its structure. In a building, the behavior relates, for instance, to what can be achieved as a result of using the building, such as (i) mild temperatures due to the material used to build it, (ii) walkable spaces, or (iii) commercial rooms that enable enterprises to base themselves and perform their businesses.

When we transplant that notion to the systems (and software) engineering world, we also need to map those concepts accordingly. We discuss the notion of system itself here; but we state that, for the purposes of this book, every system (of any nature, i.e., an electrical system, hydraulic system, a building itself, or a software-based system) has an architecture. Analogously, software-based systems architectures also have structure and behavior. When we migrate to SoS, the structure of a SoS is exactly the set of constituents that compose it and the communication links that support their interaction; while the behaviors consist of the complex functionalities that can be delivered as the conjunction of the constituents and the combination of the functionalities delivered by them. A software architecture corresponds to the fundamental structure of a software system, which comprises software elements, relations among them, and the rationale, properties, and principles governing their design and evolution (Bass et al., 2012; ISO/IEC/IEE, 2011). In turn, an architecture of a SoS is its fundamental structure, which includes its constituents and connections between them, their properties as well as those of the environment (Nielsen et al., 2015). Then, when we think about systems architectures in a smart city environment, we think about the systems that will compose each planned application (although we expect that even non-predicted constituents could also join the smart city system as it evolves), the topology of how they are linked and how their functionalities can be combined to achieve more complex functionalities. This is only possible if those systems interoperate.

As mentioned in Section 1.2, interoperation or interoperability refers to the ability of the systems to effectively communicate, exchange information, and use the information exchanged to perform actions (ISO/IEC/IEE, 2011). The interoperability links are the glue that unify the constituents and build the SoS format (Fernandes et al., 2020; Fernandes, et al., 2022). Figure 3.3 illustrates interoperability links between constituents of a smart city, represented as dotted lines. Each dotted line shows a bidirectional channel that enables the communication between the pair of systems involved. This figure illustrates that autonomous cars communicate between themselves to check if they are too close, if some of them stopped (to avoid a crash) or even to regulate the speed to keep all of them at a fixed speed. One of the autonomous cars converse with the smart traffic control and also with the smart parking system, so that it can be safe in the traffic and also find a place to park.

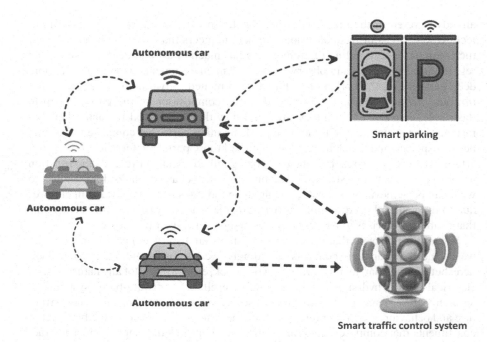

FIGURE 3.3 Example of interoperability links between constituents in a smart city.

Achieving that effective communication between the involved constituents often requires the adaptation of interfaces, protocols, and standards to enable bridging between legacy and newly designed systems (Nielsen et al., 2015). This is necessary because, in general, systems from different companies often use different forms to represent information, such as the inherent divergences of different operating systems (MacOS, Windows, or Linux), use of different technologies for communication (Wi-Fi, infrared, or 3G/4G/5G), or different protocols or technologies (REST, JSON, or XML). Even in a single company, this can happen, such as in the case of two different machines using two versions of Windows: 32 or 64 bits; this is abstracted, but a sort of mapping is still needed. The integration among systems is a recurring task in modern companies, such as the use of a chatbot in WhatsApp: they are different systems (the chatbot and the WhatsApp), but they communicate in some way to offer the experience of using the company system in the WhatsApp and the user feels as using a single system, although WhatsApp was designed exclusively for communication.

Strategies then need to be developed or used to establish communication links between those heterogeneous systems. For instance, if we have two mobiles from different brands (an iPhone and a Galaxy), they use different operational systems; however, they can communicate by means of a common application installed, such as WhatsApp or using a common protocol or technology. After that discussion, it is possible to understand that the architecture of a smart city relies on the establishment of interoperability links that abstract, that is, hide or make it transparent (not perceivable), the heterogeneity among the involved constituents and enable them to interoperate in a dynamic environment.

The combination of the five characteristics discussed in Section 3.2.2 makes SoS to be naturally *dynamic at runtime* with regard to their constituents, environment, and changing missions. Hence, the main difference between an SoS and a non-SoS large system is the nature of their constituent systems, specifically regarding their autonomy/independence level. SoS also differ from other complex networked systems, since all SoS can be characterized as distributed systems (i.e., independent systems communicating by means of some network), but not all these systems are SoS. Moreover, software has an essential influence on the design, construction, deployment, and evolution of SoS and their constituents.

3.2.6 SoS Sub-Types and Taxonomy

Other classes of SoS have also been mentioned in the literature, such as the cyber-physical systems of systems (CPSoS) (Bondavalli et al., 2016), systems-of-information systems (SoIS) (Saleh and Abel, 2015; Graciano Neto et al., 2017b; Graciano Neto et al., 2017; Fernandes et al., 2018; Teixeira et al., 2019a; Fernandes et al., 2019; Fernandes et al., 2020; Fernandes et al., 2022), and systems-of-autonomous systems (SoAS). The first type comprises those SoS whose constituents are cyber-physical systems (CPS), that is, digital systems that have some physical counterpart that senses and acts on the environment. Smart sensors are examples of CPS. The second type comprises the SoS formed by (not exclusively, but necessarily with some) information systems (IS), that is, software-based systems that are concerned with information processing and storage and whose information often supports decisions in organizational environments. And the last type is when the involved constituents have AI and machine learning abilities, continuously learning new behaviors, and constantly evolving the entire SoS as a consequence (Torkjazi and Raz, 2022). The definition of IS will not be deeply discussed herein, but it is interesting to observe that smart cities are closer to CPSoS than to SoIS. CPSoS have necessarily the physical part and act in the open world, with interoperability among those physical (but digital) systems, while SoIS regard the alliances among different companies and their IS to achieve business goals and strategies.

The inherent characteristics of SoS make its software architecture an essential element to be addressed during their development, operation, and evolution. Software architecture can support the understanding of the structure of these systems and makes it possible to deal with emergent behavior. It also incorporates quality attributes for SoS, including interoperability, reliability, adaptability, and scalability.

SoS have been largely distinguished by the level of managerial control (Dahmann et al., 2008), and by their central authority and collaboration level (Bryans et al., 2013; Pérez et al., 2013). The U.S. DoD makes use of four categories of SoS (US Department of Defense, 2008; Nielsen et al., 2015). According to Nielsen et al. (2015), three of these categories were originally defined by Maier (1998), while the "acknowledged" type was later proposed by Dahmann et al. (2009). Such categorization is required to guide the selection of architectural principles. These are the essential types of SoS and their respective characteristics (Dahmann et al., 2009; Lane, 2013; Maier, 1998):

- **Directed:** Those in which the integrated SoS is built and managed to fulfill specific purposes. It is centrally managed during long-term operation to fulfill those purposes, and any new ones the system owners may wish to address. Constituent systems maintain an ability to operate independently, but their normal operational mode is subordinated to the central managed purpose. For example, most integrated air defense networks are centrally managed to defend a region against enemy systems, although its constituents retain the ability to operate independently, and do so when needed under the stress of combat. In short, their systems operate subordinated to the central managed purpose (Pérez et al., 2013).
- **Acknowledged:** These SoS have capability objectives, management, and resources to support the SoS and are composed of existing systems along with new developments. This type of SoS has recognized objectives, a designated manager, and resources for the SoS. However, the constituent systems retain their independent ownership, objectives, funding, and development and sustainment approaches. Changes in the systems are based on collaboration between the SoS and the system (Dahmann et al., 2008). In short, there exists a considerable independence of the constituent systems, but there are specific objectives, management team, and resources addressed.
- **Collaborative:** Collaborative SoS are distinct from directed systems since the central management software entity does not have coercive power to run the system. Constituent systems must, more or less, voluntarily collaborate to fulfill the agreed upon central purposes. The Internet is a collaborative system. While standards are defined to it, there is no absolute power to enforce them. Agreements among the central players on service provision and rejection provide what enforcement mechanism there is to maintain standards. The Internet began as a directed system, controlled by the U.S. Advanced Research Projects Agency, to share computer resources. Over time, it has evolved from central control through unplanned collaborative mechanisms (Maier, 1998).
- **Virtual:** Virtual SoS lack both a central management authority and centrally agreed upon purposes. Large-scale behavior emerges and may be desirable, but the whole system must rely upon relatively invisible mechanisms to maintain it. Constituents are not subjected to a central management authority and there is not a clear SoS purpose. Constituent systems are not necessarily known and the mechanisms to maintain the SoS are not evident. Some examples are the current form of the World Wide Web and national economies. Both systems are distributed physically and managerially. The World Wide Web is even more distributed than the Internet since no one plays a central control, except at the earliest stages (Lane, 2013; Maier, 1998). However, some SoS theorists question whether this type really exists.

3.3 EXAMPLES OF SYSTEMS OF SYSTEMS

3.3.1 SoS in Smart Cities

Smart cities are a prime example of a SoS. They are a collection of interconnected systems that work together to improve the quality of life for citizens. These systems are often connected through a common network and use data from one another to make decisions and improve overall efficiency. Here is a detailed section on examples of SoS in smart cities:

- **Transportation:** Smart cities use transportation systems such as ITS, also known as STCS, to improve mobility and reduce traffic congestion. ITS can include systems for traffic management, public transportation, and connected vehicles. These systems often use real-time data from sensors and cameras to optimize traffic flow and reduce travel time. Examples of existing applications include the CityMobil2 project in Lausanne, Switzerland and the Connected Vehicle Pilot Deployment Program in the United States.
- **Energy management:** Smart cities use energy management systems to optimize the use of energy resources and reduce energy consumption. These systems can include systems for smart grid, energy storage, and building automation. These systems often use real-time data from energy meters and sensors to optimize energy usage and reduce costs. An example of this is the Amsterdam Smart City project,[3] which aimed to make Amsterdam more sustainable by using smart energy systems to optimize energy usage and reduce costs.
- **Public safety:** Smart cities use public safety systems to improve the safety and security of citizens. These systems can include systems for emergency response, surveillance, and crime prevention. These systems often use real-time data from cameras and sensors to improve response times and reduce crime. An example of this is the Safe City project in London, UK, which uses CCTV cameras and other sensors to improve public safety and reduce crime.
- **Environmental monitoring:** Smart cities use environmental monitoring systems to improve the quality of the environment. These systems can include systems for air and water quality monitoring, waste management, and weather forecasting. These systems often use real-time data from sensors and cameras to improve the quality of the environment and reduce pollution. An example of this is the EIT Climate-KIC Smart Cities project in Europe, which aimed to develop sustainable, low-carbon cities.
- **Governance:** Smart cities use governance systems to improve the efficiency and transparency of government services. These systems can include systems for e-government, citizen engagement, and data analytics. These systems often use real-time data from various sources to improve the efficiency and transparency of government services. An example of this is the Smart Dubai project in Dubai, which aims to make Dubai the smartest and most sustainable city in the world by using technology to improve government services and citizen engagement.

- **Smart parking:** Smart parking systems use a combination of sensors, cameras, and real-time data to optimize the use of parking spaces and reduce traffic congestion. These systems can include systems for parking space occupancy detection, parking guidance, and parking reservation. An example of this is the Smart Parking project in Singapore, which uses sensors, cameras, and real-time data to optimize the use of parking spaces and reduce traffic congestion.
- **Smart lighting:** Smart lighting systems use a combination of sensors, cameras, and real-time data to optimize the use of street lighting and reduce energy consumption. These systems can include systems for lighting control, energy management, and public safety. An example of this is the Smart Lighting project in Barcelona, Spain, which uses sensors, cameras, and real-time data in lampposts to optimize the use of street lighting and reduce energy consumption.
- **Smart water management:** Smart water management systems use a combination of sensors, cameras, and real-time data to optimize the use of water resources and reduce water consumption. These systems can include systems for water supply, drainage, and sewage treatment. An example of this is the Smart Water Management project in Amsterdam, Netherlands, which uses sensors, cameras, and real-time data to optimize the use of water resources and reduce water consumption.
- **Smart waste management:** Smart waste management systems use a combination of sensors, cameras, and real-time data to optimize the collection and disposal of waste. These systems can include systems for waste collection, recycling, and composting. An example of this is the Smart Waste Management project in San Francisco, USA, which uses sensors, cameras, and real-time data to optimize the collection and disposal of waste.
- **Smart building:** Smart building systems use a combination of sensors, cameras, and real-time data to optimize the use of buildings and reduce energy consumption. These systems can include systems for building automation, energy management, and security. An example of this is the Smart Building project in Frankfurt, Germany, which uses sensors, cameras, and real-time data to optimize the use of buildings and reduce energy consumption.

These examples illustrate how SoS in smart cities can bring many benefits such as increased efficiency, improved decision-making, and new capabilities. The integration of different systems, technologies, and services can lead to a better quality of life for citizens, more sustainable cities, and a more efficient use of resources.

3.3.2 OTHER APPLICATION OF SOS

Other SoS exist in environments that are not necessarily related to smart cities. For instance, a known SoS is Global Earth Observation System of Systems (GEOSS). It comprises a set of systems combined to provide operational capabilities from space, such as to monitor ocean conditions, take pictures of the Amazon region, and disseminate climatic and/or environmental information such as hurricanes. It is important to reinforce that the constellation of satellites could even provide other services such as satellite Internet. The systems that contribute to GEOSS come from the participating countries (Butterfield et al., 2008).

In general, space systems are SoS. A space SoS is composed of constituents in ground and space that accomplish missions, such as telecommunication, global position (GPS), weather forecast, Earth and space observation, meteorology, resource monitoring, and military observation. Space SoS can contain even more than 800 constituents (Yamaguti et al., 2009). Space systems are usually classified into three main segments, namely (Rohling et al., 2019): (i) *spatial*, which is the part placed in orbit (satellites, space probes, and space stations); (ii) *launcher*, used for placing the space instruments and constituents in orbit (rockets and space shuttles); and (iii) *ground*, which supervises satellite operations. Ground consists of a mission control system, an operation control system, ground stations, and data communication networks (Wertz and Larson, 1999; ECSS, 2008). Each segment materializes one or more systems that have their attributions. In systems engineering, missions are denied and constituents are articulated for a certain space mission. Satellites are the main constituents of a space SoS. These systems interoperate to offer several important services. They were not necessarily designed to be part of a SoS, but, eventually, their functionalities can be combined to obtain non-predicted behaviors. The general mission of such space SoS is *data collection*. The several types of data collection include water and deforestation index, which can be accessed online. This SoS was designed to undertake two missions, namely (i) taking pictures (monitor) of the Amazon region and (ii) distribution of environmental data collected by Data Collection Platforms (DCP) (Graciano Neto et al., 2018b).

Defense systems, in general, are also SoS. Paes et al. (2019) presented, through an experience report, SisGAAz, a SoS developed and managed by the Brazilian Navy for the purposes of maritime national security, making possible continuous monitoring and control, assuring national sovereignty, and protection. Consequently, SisGAAz involves a diversity of interdependent systems that jointly provide a large infrastructure that enables the emergence of the behaviors that can assure Brazilian national sovereignty.

3.4   DEVELOPING SYSTEMS OF SYSTEMS

Systems engineering processes have been extended to SoSE, creating such a new area of knowledge disseminated as SoSE. SoSE has its processes, methods, and techniques. Many authors have described and proposed processes to accomplish SoS conception (Butterfield et al., 2008; US Department of Defense, 2008; Farcas et al., 2010; INCOSE, 2012; Kazman et al., 2013; Walden, 2007). These activities may be conducted by members of single or multiple SoSE teams depending on the size or scope of the SoS.

One possible perspective to understand SoSE is that, at the end, SoS are engineered to realize emergent behaviors. Under that perspective, the engineer has two options to engineer SoS (Teixeira et al., 2020; Silva et al., 2022).

In a *top-down* approach, given a set of established complex goals assigned to the SoS, the engineer refines each complex goal, breaking it down into smaller operational pieces until they reach a granularity that matches the granularity of the functionalities offered by the available constituents. Then, she/he possibly combines the constituents in a data flow (or a business process) that interoperates them and enables the realization of the intended behaviors as the result of the combination of their individual capabilities.

In a *bottom-up* approach, the engineer observes the constituents available (and their respective functionalities) to be used in the design of emergent behaviors and creatively combine their features by interoperating the constituents, creating data flows until the set of constituents can offer the desired results, that is, the planned goals realized into emergent behaviors. These perspectives are constructed during the design phase of a SoS and might be supported by AI mechanisms.

In most cases, the technical requirements for a system or a system increment have been defined and are provided to the systems engineer as a starting point. In SoS, since requirements can be at a higher level, come from different sources, or in terms of capabilities rather than requirements, the systems engineer has an important role, working with stakeholders and with the SoS manager to articulate the high-level technical SoS requirements that will provide a basis for the SoSE.

Similarly, identifying the systems affecting SoS objectives and understanding their technical and organizational relationships goes beyond what is typically done by the systems engineer to address the interfaces for a new system. Finally and most importantly, the SoS engineer should pay considerable attention and invest substantial time understanding changes that are outside his or her span of control but could potentially impact the SoS. The SoSE team monitors these influences and assesses feedback on the SoS from the field as well as the results of other core elements. The SoS engineer focuses on understanding and, in fact, anticipating change as a core element of the SoSE. A central role of SoSE is the establishment and maintenance of a persistent technical framework to guide SoS evolution through the development of an evolving SoS architecture. The SoS architecture provides an integrated view of the ensemble of systems within the SoS. The development of the architecture of an SoS is an important core element for SoSE because it frames and supports design changes to the SoS over time (US Department of Defense, 2008).

On the other hand, the International Council on Systems Engineering (INCOSE, 2012) believes that a detailed SE process applied during the entire life cycle of an SoS should be tailored and expressed in terms of the ISO/IEC 15288:2008 (ISO/IEC/IEEE, 2008) processes and their outcomes, relationships, and sequence, and comprises the following steps: (i) exploratory research (which identifies stakeholders' needs, and explore ideas and technologies to conceive a system); (ii) conceptualization (which refines stakeholders' needs, explore feasible concepts, and propose feasible solutions); (iii) development (which refines system requirements, create solution description, specify the system, and verify and validate system); (iv) production (which produces systems, inspect, and verify them); (v) utilization (which happens as long as systems in execution, being operated to satisfy users' needs); (vi) support (which provides sustained system capabilities); and (vii) retirement (when a system becomes obsolete, and it is stored, archived, or discarded). Other standards have been established over the years, such as ISO/IEC/IEEE 21840:2019, which provides guidance on the application of processes in ISO/IEC/IEEE 15288 to SoS.

Walden's process is based on a classical V-model, and it is similar to software engineering processes as a whole (Walden, 2007). SoSE engineering, under that perspective, receives technologies (especially commercial-off-the-shelf or simply COTS) as input. Butterfield et al. (2008) provide a complete, detailed, and systematic SoSE process that can be applied in military and civil SoS. They consider that, in SoS, an integrator role is responsible for bringing together individual systems. They discuss a SoSE process that adopts a disciplined, enterprise wide sort to integrate systems, operations, and information-based tools and processes, and how the integration of disparate technologies provides a unique product that creates system synergies across complementary interfaces. They work on a model-based approach, that is, they use models to represent every artifact produced within the process, but in fact, integration is still performed in a non-automatized way, but in an *ad-hoc* modality. This is not an exhaustive list of processes and others can be found in the specialized literature; particularly, some other processes are mentioned in the work by Paes et al., 2019.

3.4.1 A Practical Example of SoSE in Smart City Context

To make it clear how to develop a SoS, let us use an example. Consider a smart traffic system, which should support the following goal: *a road safe crossing for a pedestrian*. This is evidently only one goal among many needs that should be support by a smart traffic system, such as (i) avoid collisions, (ii) provide a fluid traffic in a fixed speed, and (iii) monitoring the traffic laws. To achieve the established goal, we can follow an adapted version of the INCOSE process, as follows:

i. **Exploratory research:** In this step, we identify the stakeholders' needs, and explore ideas and technologies to conceive a system. This is already established: our aim is to support that pedestrians can cross a street safely and that the systems involved can contribute to reinforce the safety quality attribute.

ii. **Conceptualization:** Herein, we refine stakeholders' needs, explore feasible concepts, and propose feasible solutions. This can be performed in a top-down or bottom-up approach. In general, bottom-up approach makes sense when we want to creatively explore the available systems. Here we have the opposite situation: we have a pre-established goal and we know the systems involved. Then, we need to understand how to operationalize it. Then, we decide to follow a top-down approach. To achieve that, we think about the systems involved: autonomous cars, traffic light system, and a smart crosswalk with sensors. This can make up a feasible solution to the stated problem.

iii. **Development:** In this step, engineers refine system requirements, create solution description, specify the system, and verify and validate the SoS. To refine the requirements and create the solution description, let's use the idea

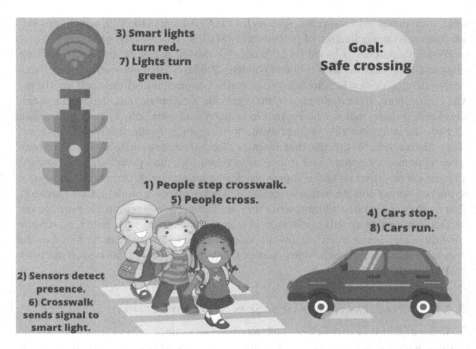

FIGURE 3.4   An illustration of a smart traffic system SoS achieving the goal "safe crossing."

of workflow to understand the problem. To make the pedestrian cross the street safely, the involved systems should accomplish the following steps, as summarized in Figure 3.4:

1. The person steps the crosswalk. When this happens, the sensors available in the crosswalk detect the presence and send a request of stop for the smart traffic light system.

2. In parallel, the autonomous cars are also equipped with depth cameras, computational vision, AI, and sensors, so that if a person appears in front of the car, in case the crosswalk sensors are disabled, the cars can detect the presence and avoid the person is ran over (as happened in the case described in the following gray box). As the person's presence is detected, they send a signal to the traffic light system.

3. The traffic light system receives stop request signals from cars and crosswalk. Any of them could trigger the lights to turn red. As this happens, the smart traffic light system notifies back the cars that slow down to stop. The cross permission sign is turned green to the person.

4. Person crosses. When she/he runs out of the crosswalk, the sensors system notifies the traffic light system, which turns green and notifies the cars, so they can go ahead to their destinations.

The specification itself could be done using a simulation model, as it will be discussed in Chapter 4, or using SosADL, a language for SoS architectural description (for further reading on it, check Oquendo (2016)). For the purposes of this small example, let us consider the specification is ready. After that, the solution can be submitted to the validation and verification of an expert (a traffic engineer, for instance) and the SoS can then be built, which is the next step:

- iv. **Production:** In this step, a team of engineers can build or acquire the involved systems, inspect, and verify them, and establish the interoperability links among them.
- v. **Utilization:** This is expected to be the longer step of the process, which happens as long as systems are in execution, being operated to satisfy users' needs.
- vi. **Support:** This is necessary to sustain the SoS operation, potentially correcting errors, replacing sensors and pieces, or improving the system software.
- vii. **Retirement:** When a system becomes obsolete, and it is stored, archived, or discarded.

A FATAL CRASH

A Uber's self-driving car ran over a pedestrian, causing a fatal crash. The car was travelling at 39 mph (63 km/h). The findings raised a series of safety issues but did not determine the probable cause of the accident. The fatal crash occurred in March 2018, and involved a Volvo XC90 that Uber had been using to test its self-driving technology. Just before the crash, Ms Herzberg had been walking with a bicycle across a poorly lit stretch of a multi-lane road. According to the U.S. National Transportation Safety Board (NTSB), Uber's test vehicle failed to correctly identify the bicycle as an imminent collision until just before impact. By that time, it was too late for the vehicle to avoid the crash. "The system design did not include a consideration for jaywalking pedestrians," the NTSB said. The report also said there were 37 crashes of Uber vehicles in self-driving mode between September 2016 and March 2018. This news remark how important is that all the systems involved in a smart traffic SoS, as that illustrated in Figure 3.4, should reliably work, since failures or lack of precision in any of them could cause serious injuries and losses as those reported by BBC.

Read more at: <https://www.bbc.com/news/technology-54175359>

3.4.2 REFERENCE ARCHITECTURES

In the realm of software-intensive SoSE, specifically, a process was created to establish, model, and validate missions of SoS by means of an artifact known as reference architectures, proposed by Garcés and Nakagawa (2017). A reference architecture is an artifact, often captured in a set of models, which is abstract enough to enable one to derive a family of several conformant architectures. In their work, Garcés and

Nakagawa (2017) establish a process, that is, a set of well-defined and interdependent steps, to establish, model, and validate missions of SoS. In that process, the conception of a SoS is driven by the establishment of missions and how constituents can be selected with their functionalities to match the functionalities. However, this is done still at reference architecture level, which allows one to derive several SoS architectures that can be adherent to that reference architecture.

3.5 CHALLENGES

Based on the state of the art of SoS and implications to the smart cities domain, some challenges can be pointed out for the next few years. Such challenges are defined based on key areas and are presented as follows:

- **Interoperability:** Researchers and practitioners still need some support on how to model and analyze a SoS based on interoperability approaches that better fit the desirable characteristics of such types of arrangements. Although interoperability has been broadly investigated by the Information Systems Community in the past five decades (Fernandes et al., 2018), establishing interoperability links in SoS to elaborate a set of implementation recommendations should cover the technical, human, and business factors altogether. Especially in the smart systems domain, the diversity of technologies, applications, and stakeholders require information visualization approaches to support strategies for making interoperability somehow smooth among the existing and future constituents of a SoS, given its dynamic architecture and evolution (Fernandes et al., 2022).
- **Reliability:** The dynamism and uncertainties in a SoS requires access to up-to-date data from its constituents, but it is challenging since they can be seen as black box elements. To cope with this situation, directed SoS use to be the most common type of SoS explored in academic and real case studies, since studies that bring solutions for sharing data from constituents are missing (Ferreira et al., 2022). Moreover, constituents could be replicated to assure that any SoS need is addressed based on redundancy, mainly when a given constituent fails, which is critical in the smart cities domain. The implementation of such a strategy may be complicated and may produce operating and maintenance costs. A final challenge in this subject refers to the selection of the most appropriate constituents to compose a SoS as a strategy to know failures prior to its operation. In this case, algorithms should consider attributes to identify limitations and restrictions of potential SoS, for example, interoperability, scalability, and performance (Ferreira et al., 2022).
- **Security:** It is worth highlighting that SoS managers and architects face difficulties in coping with information security in both the whole (a SoS) and its parts (constituents), because of vulnerabilities to threats and impacts caused by cyberattacks (Olivero et al., 2022). Based on academic studies and experts' opinions, a grand challenge in this regard consists of the fact that stakeholders need to better know the vulnerabilities, exposure,

and contribution technology makes to prevent cyberattacks and mitigate SoS risks (Dias et al., 2022). This is especially relevant when different organizations – public and private – join themselves to provide different solutions (mostly constituents) to a smart city SoS. In this context, specific methods and tools for the identification of vulnerability points and strategies to implement security policies can be indicated as research and practice challenges.

- **Process:** As previously mentioned in this chapter, a remarkable example of a SoS is a smart city whose purpose is the provision of smart services to citizens based on the collaboration of multiple entities (i.e., constituents and stakeholders) from an information flow that takes into account different business people, organizations/processes, and technologies (Fernandes et al., 2020). This process-based perspective has also been investigated as a class of SoS known as SoIS (Fernandes et al., 2018). Regardless of the way the community has named such systems arrangement, the specification and implementation of processes to support the SoSE based on different factors (technical, human, and business) are still a challenge for a mature research and development of practical solutions in industry. Exploring a systems thinking perspective should support a systemic understanding of the constituents relationships and how they properly addressed stakeholders' and organizations' needs (Cordeiro et al., 2020).

- **Architecture:** Challenges regarding architecture have been largely pointed out by the SoS research community and comprise the design, documentation, and analysis of traditional systems and architectures over the past 20 years. Given the notion of software reuse, methods and tools to support SoS architecture share several principles from the traditional systems but should be rather instantiated to address scalability and other critical quality attributes (Santos et al., 2014). In this context, system dynamics theory is a useful instrument to simulate several configurations of SoS constituents to improve the architectural design (Cordeiro et al., 2020). Moreover, the SoS evolution reveals that the notion of complexity can be applied to support the definition and evolution of smart cities SoS architectures, mainly because of its dynamic nature. SoS reference architectures and variability management consist of grand challenges to be addressed in this subject (Graciano Neto et al., 2020).

- **Governance:** Constituents that compose a SoS are independent and have their own goals. As such, one of the most critical issues consists of enabling SoS governance, which depends on several factors that should be considered when implementing strategies to do so (Imamura et al., 2020). Different from the traditional structures for information systems management, which have reached a high maturity level, a SoS depends on some kind of orchestration of its constituents with their managerial and operational independences. Factors related to SoS quality improvement, standardization of processes and products, regulations of constituents (entry/exit), lack of coordination and cooperation, lack of institutional policies for systems integration and interoperability, and integrated management issues are pointed out as opportunities for investigation in SoS governance.

- **Ecosystem:** Finally, the software ecosystem perspective can aid the under-standing of a SoS by exploring the existing relationships among its con-stituents as well as their nature. For example, the notion of ecosystem architecture can be applied to a SoS seen as an integrated common tech-nology platform composed of several constituents, or simply an industry platform (Santos et al., 2014). Moreover, this parallel raised the concept of system-of-information systems ecosystem (EcoSoIS), that is, a software ecosystem that involves the development and interoperable activity within a set of constituents (information systems) working together to support busi-ness and social goals (Graciano Neto et al., 2020). A critical barrier refers to the identification of how the constituents' relationships can benefit and/or harm an organization that delivers products and services. This is observed in smart cities in which a public organization (government) is mostly responsible for delegating and orchestrating the responsibilities based on different software-intensive systems (i.e., who will address which goal). In this case, the ecosystem perspective can support citizens, companies and governments to combine those assets to cooperatively work together to achieve a bigger business value. Co-innovation is a relevant challenge to be addressed in this context from social web data mining and analysis, since a smart city requires cooperation and collaboration.

3.6   FINAL REMARKS

The relevance, potential, and complexity of SoS have considerably impacted how these interconnected systems have been developed, operated, and evolved in dif-ferent application domains, including in smart cities. In this scenario, this chapter defined SoS and its main characterization of SoS, particularly from the smart cities perspective.

Diverse research and industry initiatives have been the driving force to enable SoS construction and evolution. However, it is still imperative to learn from past experiences from diverse other communities, including systems engineering, real time, and embedded systems engineering, and provide general theories and tech-nologies for mastering the complexity of SoS. Hence, conceiving rigorous founda-tions, languages, processes, and tools for supporting the engineering of SoS are the steps to mature this area, together with carrying out open discussions and exchange of knowledge and experience to address the challenges raised by SoS core charac-teristics. The next chapter deals with simulation models and how they can be used to help in engineering smart cities.

NOTES

1 Note: We clarify that, in the literature, authors use "an" before SoS in the case you read it character-by-character. In case you read it entirely, we use "a." Both forms are found herein.
2 Henceforth, the acronym "SoS" will be interchangeably used to express both forms: singular (system of systems) and plural (systems of systems).
3 https://amsterdamsmartcity.com/

REFERENCES

Alampalli, S., & Pardo, T. (2014). A study of complex systems developed through public private partnerships. In 8th International Conference on Theory and Practice of Electronic Governance (pp. 442–445). Guimaraes, Portugal.

Axelsson, J., & Nylander, S. (2018). An analysis of systems-of-systems opportunities and challenges related to mobility in smart cities. In 2018 13th Annual Conference on System of Systems Engineering (SoSE) (pp. 132–137). IEEE, Paris, France.

Baldwin, W., & Sauser, B. (2009). Modeling the characteristics of system of systems. In IEEE International Conference on System of Systems Engineering (SoSE) (pp. 1–6). Albuquerque, New Mexico.

Bass, L., Clements, P., & Kazman, R. (2012). Software Architecture in Practice. Addison-Wesley Longman Publishing Co., Inc., Boston, MA.

Basso, F. P., Kreutz, D., Molina-Jiménez, C., & Frantz, R. Z. (2022). On the use of emerging decentralised technologies for supporting software factories coopetition. International Journal of Computer Applications in Technology, 69(2), 123–138.

Boardman, J., & Sauser, B. (2006). System of Systems – the meaning of. In IEEE/SMC International Conference on System of Systems Engineering (pp. 1–6). Los Angeles, CA.

Bondavalli, A., Bouchenak, S., & Kopetz, H. (2016). Cyber-Physical Systems of Systems Foundations – a Conceptual Model and Some Derivations: The Amadeos Legacy. Springer International Publishing, Gewerbestrasse, Cham.

Bryans, J., Payne, R., Holt, J., & Perry, S. (2013). Semi-formal and formal interface specification for system of systems architecture. In 2013 IEEE International Systems Conference (SysCon) (pp. 612–619). Orlando, FL, USA.

Bulcão-Neto, R., Teixeira, P., Lebtag, B., Graciano Neto, V., Macedo, A., & Zeigler, B. (2022), Simulation of IoT-oriented fall detection systems architectures for in-home patients. IEEE Latin America Transactions, 14(8), 1–9.

Butterfield, M. L., Pearlman, J. S., & Vickroy, S. C. (2008). A system-of-systems engineering GEOSS: Architectural approach. IEEE Systems Journal, 2(3), 321–332.

Cavalcante, E., Cacho, N., Lopes, F., & Batista, T. (2017). Challenges to the development of smart city systems: A system-of-systems view. In XXXI Brazilian Symposium on Software Engineering (pp. 244–249). Fortaleza, CE, Brazil.

Cavalcante, E., Cacho, N., Lopes, F., Batista, T., & Oquendo, F. (2016). Thinking smart cities as systems-of-systems: A perspective study. In 2nd International Workshop on Smart (pp. 1–4), Trento, Italy.

Chalmers, D. J. (2006). Strong and weak emergence. In P. Davies, P. Clayton (Eds) The Re-Emergence of Emergence: The Emergentist Hypothesis From Science to Religion (pp. 244–256). Oxford University Press, Oxford.

Cordeiro, F., Vasconcelos, A., Santos, R., & Lago, P. (2020). Towards an accountability suggestion map for supporting information systems management based on systems thinking. In IEEE 21st International Conference on Information Reuse and Integration for Data Science (IRI) (pp. 295–300).

Dahmann, J., Rebovich, G., & Lane, J. (2008). Systems engineering for capabilities. CrossTalk Journal – The Journal of Defense Software Engineering, 21(11), 4–9.

Dahmann, J., Baldwin, K., & Rebovich, G. (2009). Systems of systems and net-centric enterprise systems. In 7th Annual Conference on Systems Engineering Research (pp. 1–6). Loughborough, UK.

Delécolle, A., Lima, R., Neto, V., & Buisson, J. (2020). Architectural strategy to enhance the availability quality attribute in system-of-systems architectures: A case study. In 15th International Conference of System of Systems Engineering (SoSE) (pp. 93–98). Budapest, Hungary.

Dersin, P. (2014). *Systems of Systems*. IEEE-Reliability Society. Technical Committee on Systems of Systems. https://rs.ieee.org/technical-activities/technical-committees/systems-of-systems.html (Access in 12/11/2022).

Dias, R., Zacarias, R., Varella, J., & Santos, R. (2022). Investigating information security in systems-of-systems. In XVIII Brazilian Symposium on Information Systems (pp. 1–8). Curitiba, Brazil.

Dickerson, C., & Mavris, D. (2011). Relational Oriented Systems Engineering (ROSE): Preliminary report. In 6th International Conference on System of Systems Engineering (pp. 149–154). Albuquerque, NM.

ECSS. (2008). ECSS Space Engineering – Ground systems and operations. European Cooperation for Space Standardization (ECSS). ·

Farcas, C., Farcas, E., Krueger, I., & Menarini, M. (2010). Addressing the integration challenge for avionics and automotive systems—From components to rich services. Proceedings of the IEEE, 98(4), 562–583.

Fernandes, J., Cordeiro, F., Ferreira, F., Neto, V. V. G., & dos Santos, R. P. (2022). A method for identification of potential interoperability links between information systems towards system-of-information systems. iSys-Brazilian Journal of Information Systems, 15(1), 1–26.

Fernandes, J., Ferreira, F., Cordeiro, F., Neto, V. V. G., & dos Santos, R. P. (2019). A conceptual model for systems-of-information systems. In 20th International Conference on Information Reuse and Integration for Data Science (IRI) (pp. 364–371). Los Angeles, CA.

Fernandes, J., Ferreira, F., Cordeiro, F., Neto, V. G., & Santos, R. (2020). How can interoperability approaches impact on Systems-of-Information Systems characteristics? In XVI Brazilian Symposium on Information Systems (pp. 1–8). São Bernardo do Campo, Brazil.

Fernandes, J. C., Neto, V. V. G., & Santos, R. P. D. (2018). Interoperability in systems-of-information systems: A systematic mapping study. In 17th Brazilian Symposium on Software Quality (pp. 131–140), Caxias do Sul, Brazil.

Ferreira, F., Nakagawa, E., & Santos, R. (2021). Reliability in software-intensive systems: Challenges, solutions, and future perspectives. In 47th Euromicro Conference on Software Engineering and Advanced Applications (SEAA) (pp. 54–61). Palermo, Italy.

Firesmith, D. (2010). Profiling Systems Using the Defining Characteristics of Systems of Systems (SoS). Software Engineering Institute, Carnegie Mellon University. Pittsburgh, PA.

Garcés, L., & Nakagawa, E. Y. (2017). A process to establish, model and validate missions of systems-of-systems in reference architectures. In Symposium on Applied Computing (pp. 1765–1772). Marrakech, Morocco.

Gascó-Hernandez, M. (2018). Building a smart city: Lessons from Barcelona. Communications of the ACM, 61(4), 50–57.

Graciano Neto, V. V., Horita, F., Cavalcante, E., Rohling, A., El-Hachem, J., Santos, D., & Nakagawa, E. Y. (2018c). A study on goals specification for systems-of-information systems: Design principles and a conceptual model. In Proceedings of the XIV Brazilian Symposium on Information Systems (pp. 1–8), ACM, Marrakech, Morocco.

Graciano Neto, V. V., Horita, F. E. A., Santos, R. P. D., Viana, D., Kassab, M., Manzano, W. A. E., & Nakagawa, E. Y. (2019a). Sob (save our budget): A simulation-based method for prediction of acquisition costs of constituents of A system-of-systems. Revista Brasileira de Sistemas de Informação – iSys, 12(4), 6–35.

Graciano Neto, V. V., Oquendo, F., & Nakagawa, E. Y. (2017). Smart systems-of-information systems: Foundations and an assessment model for research development. In C. Boscarioli, R. Araujo, R. S. P. Maciel (eds) GranDSI-BR: Grand Research Challenges in Information Systems in Brazil (2016-2026) (pp. 13–24). Sociedade Brasileira de Computação, Porto Alegre.

Graciano Neto, V. V., Paes, C. E., Garcés, L., Guessi, M., Manzano, W., Oquendo, F., & Nakagawa, E. Y. (2017a). Stimuli-SoS: A model-based approach to derive stimuli generators for simulations of systems-of-systems software architectures. Journal of the Brazilian Computer Society, 23(1), 1–22.

Graciano Neto, V., Santos, R., Viana, D., & Araujo, R. (2020). Towards a conceptual model to understand software ecosystems emerging from systems-of-information systems. In R. Santos, C. Maciel, J. Viterbo (eds) Software Ecosystems, Sustainability and Human Values in the Social Web (pp. 1–20). Springer, Cham.

Graciano Neto, V. V., Basso, F., dos Santos, R. P., Bakar, N. H., Kassab, M., Werner, C. & Nakagawa, E. Y. (2019b). Model-driven engineering ecosystems. In 2019 IEEE/ACM 7th International Workshop on Software Engineering for Systems-of-Systems (SESoS) and 13th Workshop on Distributed Software Development, Software Ecosystems and Systems-of-Systems (WDES) (pp. 58–61). IEEE, Montréal, Canada.

Graciano Neto, V. V., Cavalcante, E., El Hachem, J., & Santos, D. S. (2017b). On the interplay of business process modeling and missions in systems-of-information systems. In IEEE/ACM Joint 5th International Workshop on Software Engineering for Systems-of-Systems and 11th Workshop on Distributed Software Development, Software Ecosystems and Systems-of-Systems (JSOS) (pp. 72–73). Buenos Aires, Argentina.

Graciano Neto, V. V., Garcés, L., Guessi, M., Paes, C., Manzano, W., Oquendo, F., & Nakagawa, E. Y. (2018a). ASAS: An approach to support simulation of smart systems. In 51st Hawaii International Conference on System Sciences (HICSS) (pp. 5777–5786). Big Island, HIi, USA.

Graciano Neto, V. V., Manzano, W., Rohling, A. J., Ferreira, M. G. V., Volpato, T., & Nakagawa, E. Y. (2018b). Externalizing patterns for simulations in software engineering of systems-of-systems. In 33rd Annual ACM Symposium on Applied Computing (SAC) (pp. 1687–1694). Pau, France.

Horita, F. E., Rhodes, D. H., Inocêncio, T. J., & Gonzales, G. R. (2019). Building a conceptual architecture and stakeholder map of a system-of-systems for disaster monitoring and early-warning: A case study in Brazil. In XV Brazilian Symposium on Information Systems (pp. 1–8).

Hughes, D., Ueyama, J., Mendiondo, E., Matthys, N., Horré, W., Michiels, S., Huygens, C., Joosen, W., Man, K. L., & Guan, S.-U. (2011). A middleware platform to support river monitoring using wireless sensor networks. Journal of the Brazilian Computer Society, 17(2), 85–102.

Imamura, M.,; Costa, L., Pereira, B., Ferreira, F., Fontão, A., & Santos, R. (2020). Governance factors in systems-of-systems: Analysis of a Brazilian public institution. In 5th Workshop on Social, Human and Economic Aspects of Software (WASHES) (pp. 31–40). Cuiabá, Brazil.

INCOSE. (2012). Systems engineering handbook: A guide for system life cycle processes and activities.

ISO/IEC/IEEE. (2008). ISO/IEC/IEEE international standard - systems and software engineering system life cycle processes. IEEE Std 15288-2008, 1–84.

ISO/IEC/IEEE. (2011). ISO/IEC/IEEE SO/IEC/IEEE systems and software engineering – Architecture description. IEEE Std 1471-2000, 1–46.

Jamshidi, M. (2017). Systems of Systems Engineering: Principles and Applications. CRC Press, Boca Raton, FL.

Kazman, R., Schmid, K., Nielsen, C., & Klein, J. (2013). Understanding patterns for system of systems integration. In 8th International Conference on System of Systems Engineering (pp. 141–146). Maui, HI.

Lane, J. A. (2013). What is a System of Systems and Why Should I Care? University of Southern California. CA, USA.

Maier, M. (1998). Architecting principles for systems-of-systems. Systems Engineering, 1(4), 267–284.

Mittal, S., & Rainey, L. (2015). Harnessing emergence: The control and design of emergent behavior in system of systems engineering. In Conference on Summer Computer Simulation (SummerSim) (pp. 1–10). Chicago, IL.

Nielsen, C., Larsen, P., Fitzgerald, J., Woodcock, J., & Peleska, J. (2015). Systems of systems engineering: Basic concepts, model-based techniques, and research directions. ACM Computing Surveys, 48(2), 1–41.

Olivero, M., Bertolino, A., Dominguez-Mayo, F., Matteucci, I., & Escalona, M. (2022). A Delphi study to recognize and assess systems of systems vulnerabilities. Information and Software Technology, 146(C), 1–17.

Oquendo, F. (2016). Formally describing the software architecture of systems-of-systems with SosADL. In 2016 11th System of Systems Engineering Conference (SoSE) (pp. 1–6). IEEE, Kongsberg, Norway.

Paes, C. E., Neto, V. V. G., Moreira, T., & Nakagawa, E. Y. (2019). Conceptualization of a system-of-systems in the defense domain: An experience report in the Brazilian scenario. IEEE Systems Journal, 13(3), 2098–2107.

Payne, B., Ling, L. O., & Gorod, A. (2020). Towards a governance dashboard for smart cities initiatives: A system of systems approach. In 2020 IEEE 15th International Conference of System of Systems Engineering (SoSE) (pp. 587–592). IEEE, Budapest, Hungary.

Pérez, J., Díaz, J., Garbajosa, J., Yagüe, A., Gonzalez, E., & Lopez-Perea, M. (2013). Large-scale smart grids as system of systems. In First International Workshop on Software Engineering for Systems-of-Systems (pp. 38–42). Montpellier, France.

Rohling, A. J., Neto, V. V. G., Ferreira, M. G. V., Dos Santos, W. A., & Nakagawa, E. Y. (2019). A reference architecture for satellite control systems. Innovations in Systems and Software Engineering, 15, 139–153.

Saleh, M., & Abel, M. H. (2015). Information systems: Towards a system of information systems. In KMIS 7th International Conference on Knowledge Management and Information Sharing (pp. 193–200), Lisbon, Portugal.

Santos, R., Gonçalves, M., Nakagawa, E., & Werner, C. (2014). On the relations between systems-of-systems and software ecosystems. In VIII Workshop on Distributed Software Development, Software Ecosystems and Systems-of-Systems (WDES) (pp. 58–62). Maceió, Brazil.

Santos, D., Oliveira, B., Kazman, R., & Nakagawa, E. (2022). Evaluation of systems-of-systems software architectures: State of the art and future perspectives. ACM Computing Surveys, 55(4), 1–35.

Silva, E., Batista, T., & Oquendo, F. (2015). A mission-oriented approach for designing system-of-systems. In 2015 10th System of Systems Engineering Conference (SoSE) (pp. 346–351). IEEE, San Antonio, TX.

Silva, K. C., Horita, F., & Graciano Neto, V. V. (2022). Bring us MacGyver predictor: Towards a deep learning-based mechanism to design emergent behaviors in systems-of-systems. In XXVI Brazilian Symposium on Software Engineering (pp. 299–304). Virtual Event, Brazil.

Teixeira, P. G., Lebtag, B. G. A., de Oliveira, L. W., de Carvalho, S. T., Veiga, E. F., & de Sousa Rocha, C. (2019b). Modeling and simulation of a smart street lighting system. In I Workshop em Modelagem e Simulação de Sistemas Intensivos em Software (pp. 44–48). Salvador, Brazil.

Teixeira, P. G., Lebtag, B. G. A., Dos Santos, R. P., Fernandes, J., Mohsin, A., Kassab, M., & Neto, V. V. G. (2020). Constituent system design: A software architecture approach. In 2020 IEEE International Conference on Software Architecture Companion (ICSA-C) (pp. 218–225). IEEE, Salvador, Brazil.

Teixeira, P. G., Lopes, V. H. L., Dos Santos, R. P., Kassab, M., & Neto, V. V. G. (2019a). The status quo of systems-of-information systems. In IEEE/ACM 7th International Workshop on Software Engineering for Systems-of-Systems (SESoS) and 13th Workshop on Distributed Software Development, Software Ecosystems and Systems-of-Systems (WDES) (pp. 34–41). Montréal, Canadá.

Torkjazi, M., & Raz, A. K. (2022). A taxonomy for system of autonomous systems. In 2022 17th Annual System of Systems Engineering Conference (SoSE) (pp. 198–203). IEEE, Rochester, NY.

US Department of Defense. (2008). *System engineering guide for system-of-systems engineering*. US Department of Defense (DoD).

Walden, D. (2007). The changing role of the systems engineer in a system of systems (SoS) environment. In 1st Annual IEEE Systems Conference (pp. 1–6). Honolulu, HI.

Wertz, J., & Larson, W. (1999). Space Mission Analysis and Design. Microcosm Publishing, Cleveland, OH.

Yamaguti, W., Orlando, V. & Pereira, S. D. P. (2009). Sistema brasileiro de coleta de dados ambientais: Status e planos futuros. In Simpósio Brasileiro de Sensoriamento Remoto (14, pp. 1633–1640), Natal, Brazil.

4 Modeling and Simulation for Smart City Development

4.1 INTRODUCTION

As shown in Chapter 3, smart cities are systems of systems (SoS), that is, a set of independent systems that together can deliver broader functionalities. The novel functionalities are a combination of individual functionalities offered by the constituent systems. Similarly, as mentioned earlier, SoS have dynamic/evolutionary architecture, which means that constituents can join or leave the SoS at runtime. The problem is that, if a pivotal constituent leaves the SoS at runtime, this can bring serious problems to sustain the emergent behaviors being delivered as this can disrupt the smooth functioning and stability of the entire system. The departure of a constituent can cause imbalances and negatively impact the interdependent relationships between the components, leading to unexpected behavior and potential failures. This can compromise the performance, security, and reliability of the system and cause significant consequences to the system's overall objectives.

Since SoS support critical systems, eventual failures can cause financial damage, accidents, and other serious injuries to the users. Then, it is paramount to assure, still at design-time, the consequences of changes made at the SoS architecture at runtime and actions to mitigate eventual problems. Design-time assurance refers to the process of ensuring that a system or product meets its specified requirements and adheres to certain standards at the design stage, before it is implemented or constructed. This can include reviews, inspections, and testing to verify that the design is complete, consistent, and meets functional and non-functional requirements such as security, performance, and usability. The goal of design-time assurance is to identify and correct any errors or defects early in the development process, reducing the risk of costly and time-consuming rework later on.

Simulation can then help in planning the systems and how they will be combined to form the smart cities, the behavior and structure of their combination and the impact of changes over its architecture. This chapter addresses simulation, a consolidated technology that represents the state of the art in systems engineering to provide reliability for SoS.

4.2 WHAT IS SIMULATION?

Simulation is the imitation of the operation of a real-world process or system over time (Banks, 1999). The process of simulating something typically involves creating a model of the system or process of the real world being represented, and then using that model to predict how the system or process would behave under different conditions. Simulation allows to observe how the system behaves over time, evaluating how inputs and parameters affect the outcomes and allowing predictions under different conditions. It is a powerful (and essential) tool that allows us to study the behavior (and structure) of systems that are too complex, large, or expensive to study in the real world, or that can't be studied at all by other means, also it allows to test different hypotheses and policies in a safe and controlled environment.

Simulation is particularly interesting when (i) the domain is critical and risky, which turns better to simulate than delivering a system and suffer the consequences, (ii) the system being studied is complex and a simulation can help to make it more understandable, (iii) the system is large and a simulation can help humans to study its structure and behavior, and/or (iv) the system is dynamic, that is, it can change parts of it over time, which can bring serious consequences and the resulting problems should be avoided.

Simulation can be used in a variety of fields, including engineering, physics, computer science, and finance. For example, engineers might use simulation to model the behavior of a new airplane design, while physicists might use it to model the behavior of subatomic particles. In computer science, simulation is often used to model the behavior of computer systems, networks, and algorithms. In finance, simulation is used to model the performance of investment portfolios or the risk of financial instruments.

The Modeling and Simulation Body of Knowledge (MSBoK) is a comprehensive framework that defines the knowledge, skills, and practices that are necessary for professionals working in the field of modeling and simulation (M&S). It was developed as a cooperation of more than 50 researchers over the world, and aims to provide a clear understanding of the various components of the field, as well as the relationships between them (Ören et al., 2023).

The MSBoK can be used by professionals in the field to improve their skills and knowledge, and by organizations to evaluate and improve their M&S capabilities. Additionally, it also can be used as a basis for education, training, and professional certification programs in the field of modeling and simulation.

According to the MSBoK, the three main aspects of simulation are as follows:

1. **Model development:** This aspect of simulation involves creating a mathematical or computational representation of a system or process. It includes the process of identifying the problem, defining the requirements, creating the model, and validating the model.
2. **Experimentation and analysis:** This aspect of simulation involves running simulations using the model, and then analyzing the results to draw conclusions about the behavior of the system. It includes the process of designing and executing experiments, collecting and analyzing data, and interpreting the results.
3. **Verification, validation, and accreditation:** This aspect of simulation involves the process of ensuring the fidelity and credibility of the model by comparing it with real-world data or other models. This includes the process

of verifying the model, validating the model, and accrediting the model for a specific use. Verification ensures that the model is correct, and built in accordance with the established requirements, validation ensures that the model correctly represents the system or process it is meant to simulate, and accreditation involves the formal recognition of the model's credibility and utility by a recognized body, it means that the model is fit for its intended purpose.

These three aspects work together to create an accurate and reliable model that can be used to predict and analyze the behavior of a system or process. A well-designed and executed simulation can provide insight into the system, identify potential problems, and help in decision-making.

There are many different types of simulation methods, each with its own strengths and weaknesses, and each suited to different types of systems and applications. Some of the most common types of simulation methods include following (but are not restricted to) (de França and Travassos, 2016):

1. **Discrete event simulation (DEVS):** This type of simulation models the behavior of a system as a series of events that occur over time. It is often used to model complex systems, such as manufacturing plants or transportation systems, where the order of events is important.
2. **Continuous simulation:** This type of simulation models the behavior of a system as a continuous process. It is often used to model physical systems, such as fluids and thermodynamics, where the system's state is constantly changing.
3. **System dynamics (SD):** This type of simulation is used to model dynamic systems and is based on the principles of feedback and non-linearity. It is used to study complex systems with a high degree of uncertainty, mainly in the fields of management, finance, and social sciences.
4. **Monte Carlo simulation:** This type of simulation uses random numbers to model the behavior of a system. It is often used in finance and business to model risk and uncertainty, and to make predictions about future events.
5. **Agent-based simulation (ABS):** This type of simulation models the behavior of systems made up of autonomous agents, such as individuals, firms, and animals. ABSs are used in a wide range of fields, including economics, social sciences, and engineering.
6. **Finite element method (FEM):** This type of simulation is mainly used in engineering to analyze and predict the behavior of physical systems, especially those that involve complex geometries, loads, and boundary conditions.
7. **Computational fluid dynamics (CFD):** This type of simulation involves solving mathematical equations in order to predict the behavior of fluids.

We remark that those types of simulations are not necessarily self-exclude and they can be used in association in certain contexts.

These definitions bring two important concepts: *model* and *simulator*, and these concepts should be clarified so that the reader can have a better understanding of them.

4.2.1 WHAT IS A MODEL?

Simulating requires a model (Gray and Rumpe, 2016). A model is a simplified representation of reality, and it is only an approximation of the real system. Therefore, it can contain assumptions and simplifications that can affect its accuracy and reliability. That is why it is important to validate the model and to verify it against real-world data. Additionally, it is important to remember that a model is always just a representation of a system, it cannot be the system itself (it can, however, be linked to the real system, as it is discussed in Chapter 5 on digital twins).

In the context of simulation, a model is a mathematical or computational representation of a system or process. The model captures the essential characteristics of the system or process and is used to analyze and predict its behavior. There are many different types of models and each suited to different types of systems and applications.

A model can be thought of as an abstraction of the real world, and it can take many different forms, such as equations, diagrams, computer programs, or physical prototypes. There are many different types of models, and they can be used in a wide variety of fields, such as engineering, science, finance, and business. Here are a few examples of different types of models:

1. **Mathematical models:** This type of model represents a system or process using mathematical equations. Examples include following:
   - A simple mass-spring-damper model used to model the behavior of mechanical systems.
   - A population growth model that predicts the growth of a population based on birth and death rates.
   - A weather forecasting model that uses equations to predict the movement of air masses and predict the weather.
2. **Computational models:** This type of model represents a system or process using computer programs. Examples include following:
   - A traffic simulation model that predicts traffic flow on a road network.
   - A climate model that predicts the global climate and the effects of greenhouse gases.
3. **Physical models:** This type of model represents a system or process using physical prototypes. Examples include following:
   - A wind tunnel model used to test the aerodynamics of an aircraft.
   - A prototype of a bridge used to test its structural integrity.
   - A model of a house used to test its energy efficiency.
4. **Conceptual models:** This type of model represents a system or process using diagrams, maps, or other visual representations. Examples include following:
   - A flowchart that represents the process of manufacturing a product.

- A conceptual model of a computer network that shows the connections between different devices.
- A mind map representing the different components of a business strategy.

5. **Statistical models:** This type of model represents a system or process using statistical methods, such as regression analysis or time series analysis. Examples include following:
   - A model that predicts the probability of a customer buying a product based on their browsing history.
   - A model that predicts the price of a stock based on historical data and other financial indicators.
   - A model that predicts the probability of a patient developing a certain condition based on their medical history.

It can be noticed that some of the models conceived to be simulated (simulation models) have one or more of the characteristics listed earlier. DEVS models, for instance, are computational models, but they can also have mathematical and statistical elements, being classified as such, as well.

4.2.2 WHAT IS A SIMULATOR?

A simulator is a software tool that works as an execution engine to imitate/exercise/run a model that represents the structure and behavior of a system or process over time. It is built on top of a model, which is, as aforementioned, a mathematical or computational representation of the system or process. The simulator uses the model to represent the structure and predict the behavior of the system or process under different conditions, allowing the user to analyze and understand the system in ways that would be difficult or impossible to do in the real world. Simulators enable the simulation to progress over time and usually generates a *log* (as it can be seen in the inferior part of Figure 4.1), that is, a textual representation of the history of steps over time so that auditing operations can be performed or the output of the simulation can be delivered to some other software that can analyze it and provide a diagnosis over the system being represented.

This specialized software that can check simulation results are often known as model checkers. An instance of a statistical model checker is called PLASMA. Essentially, a software can be considered a simulator if it has the execution engine itself. However, to be even more effective, a simulator can offer a visual counterpart called animation. The ability to animate models can help one better understand modeled behavior. Novices and experienced developers will both benefit from the visualization of modeled behavior provided by model animators. Model animation can give quick visual feedback to novice modelers and can thus help them identify improper use of modeling constructs. Experienced modelers can use model animation to understand designs created by other developers better and faster (France and Rumpe, 2007).

Simulators, as simulation models, can also be used in various fields such as engineering, transportation, healthcare, military, finance, and many more. A simulator

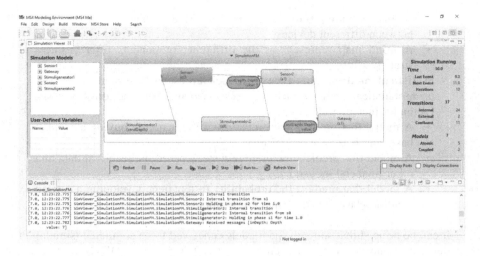

FIGURE 4.1   Screenshot of the animation provided by a MS4 Me simulator.

can take many forms. It can be a software program, a hardware device, or a combination of both. It can be used for various purposes, such as training, design optimization, testing, planning, and decision-making. It allows the user to test different scenarios in a safe and controlled environment, without affecting the real system. Different types of simulators exist and conform to the definitions brought so far. For instance, we have those simulators that can be used in driving lessons, that imitate a car, but safely prevents the learner from causing real accidents. However, it promotes the experience of driving in a faithful manner. Several sciences also provide simulators for their practice, such as chemistry (in which structures of feasible proteins can be investigated), biology (enabling to study groups behavior), and engineering, which enables them to investigate edification structures. Herein, we focus on a nonexhaustive list of simulators that (are or) can be used to study smart cities structures and behaviors, specifically.

Examples of simulators include the following:

- **Flight simulators:** Used to train pilots by imitating the behavior of an aircraft in different flight conditions
- **Process plant simulators:** Used to optimize the performance of industrial processes such as chemical and oil and gas processes
- **Traffic simulators:** Used to predict traffic flow on road networks and evaluate the impact of different traffic management strategies
- **Virtual reality simulators:** Used to train surgeons and other

Figure 4.1 shows the animation provided by a simulator named MS4 Me. MS4 Me runs DEVS models and is based on Eclipse Modeling Framework (EMF), with the interface being familiar for programmers. In that image, part of a flood monitoring system is represented. We have sensors that sense the river level and forward the collected data until a gateway that processes that information and, if necessary, triggers an alarm of flood alert. The rectangles with rounded borders represent messages being exchanged between the sensors and forwarded until the gateway. During the animation, it is possible to visualize the flow of information between the sensors, that is, which collects the information and sends it out, besides being possible to visualize the messages being explicitly transferred between the sensors. It is important to remark that simulators can only have the execution engine and log, not displaying a visual animation, and it can still be useful and considered a simulator.

In terms of software and systems engineering, we should distinguish between simulation formalism and simulator. Simulation formalism refers to the specific language, method, or technique used to create the model of the system or process being simulated. It is the set of rules and principles that govern how the model is defined and represented. For example, the mathematical equations that describe the behavior of a system, the set of computer instructions used to program a simulation and the graphical notation of a conceptual model. A simulator, on the other hand, refers to the software tool or system that is used to execute the simulation. It is the implementation of the simulation formalism. The simulator uses the model to simulate the behavior of the system over time, by calculating the state of the system at different points in time and displaying the results. A simulator can be a standalone program, a library, or a plugin; it could be a web-based, or a hardware-based.

Some of the most popular simulators that could be mentioned are MATLAB®/ Simulink, MS4 Me, PythonDEVS, System Dynamics, JADE, and NetLogo.

Simulink is a proprietary simulator of MathWorks for MATLAB formalism.[1] It supports continuous and discrete event representation and it is very popular in engineering, physics, and mathematics applications. It is based on block diagrams that can be linked to represent a complex system, but scripting (code) is also possible to complement/customize/create blocks. It provides a multitude of libraries, such as for controls, signal processing, and telecommunications applications. Then, it is often used by engineers to prototype multidisciplinary complex systems, such as a wind farm, that can involve several types of engineers.

Several applications using MATLAB for smart cities are reported in the literature. Hanumantha Rao et al. (2019) report the use of MATLAB/Simulink to simulate an electric power system for smart residential communities in smart cities. Ksiksi et al. (2015) adopt MATLAB to model an Intelligent Traffic Alert System (iTAS) that warns drivers of potential dangers on the road using audio and visual alerts. Sharma et al. (2020) report on the use of MATLAB as part of the infrastructure to model an integrated fire detection system using IoT and image processing techniques for smart cities. Chackravarthy et al. (2018) use MATLAB for numerical analysis in the context of crime anomaly detection in smart cities using deep learning.

NetLogo has a web version[2] and enables, for instance, to represent the dynamic behavior of some cars in traffic, as shown in Figure 4.2.

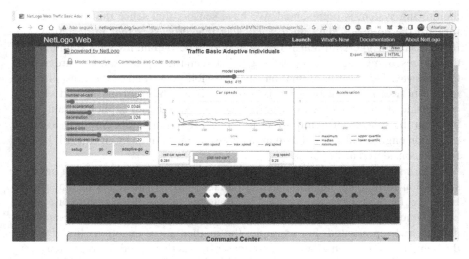

FIGURE 4.2   Traffic simulation using a NetLogo simulator.

WHAT IF WE LIVE IN A SIMULATION?

The classic movie Matrix has already brought this to our minds in the 1990s. Well, if yes, probably there is nothing we can do about it. It is unlikely that we could reach some nirvana level and, as Sophie in Sophie's World from Jostein Gaarder, simply run out of our own reality. If you search in Google for "we live in simulation," interestingly, you have strong opposite ideas, arguing against and in favor of that idea.

However, the implications of it are interesting. First, if we live in a simulation, this is certainly a really rich simulation, since the details are enormously diverse and artificial intelligence is certainly associated, since we are conscious and we even have the ability to create other simulations.

In 2003, the philosopher Nick Bostrom, from the University of Oxford, in the United Kingdom, formulated his simulation hypothesis. He argued that, in fact, it is very likely that our universe is a simulation. Physicist Seth Lloyd, from the Massachusetts Institute of Technology (MIT), in the United States, took the simulation hypothesis to the next level, suggesting that the entire universe could be a giant quantum computer. And in 2016, entrepreneur Elon Musk said that "most likely, we are in a simulation."

Overall, if we live in a simulation, it would have profound implications for our understanding of the world and the way we use simulations to design smart cities. It would require us to be more cautious in our use of simulations, and to consider the ethical implications of our actions within a simulated reality. However, as you saw, this answer is not simple. Then, let us go ahead with our lives!!

4.3   HOW SIMULATION CAN HELP ON ENGINEERING SMART CITIES?

The previous section was intended to explain what simulation is, shows examples, and types of simulation. This section shows how simulation can help to design smart cities, detailing applications, and how simulation helped in each scenario.

Before detailing specific simulators, maybe it is important to remark that simulation models are a specific type of executable models (ExM), that is, systems specifications that can, themselves, be executed so that it is possible to analyze their structures and behaviors, predicting properties, and supporting decisions on how to engineer each part of a smart city (Hojaji et al. 2019; Lebtag et al., 2021). ExM offer the ability to (i) benchmark the diverse arrangements a complex system can assume, (ii) predict system structure and behavior, despite its inherent dynamics and adaptivity, and (iii) predict the consequences of architectural changes that can take place on the fly over the structure and behavior of such system of interest, still at design-time. These advantages match the requirements imposed by SoS, and, as a consequence, by smart cities. Simulation models can be used to study the complexity of the problem being addressed by exploring different scenarios and observing the consequences of each decision visually. On the other hand, when the volume of entities to be simulated hinders the possibility to have a meaningful visual representation, professionals need to rely on the execution trace of the simulation models (Hojaji et al., 2019).

Each part of a smart city can require a different approach to be planned and, consequently, simulated. For instance, when discussing smart traffic control, it is important to adopt some type of simulation that enables one to visualize the effect of traffic elements on the overall traffic, such as the relative speed of cars and the interference of pedestrians and/or traffic lights on the flow of cars, as it can be seen in Figure 4.2 with NetLogo. Then, the visualization of vehicles movement and even 3D models can be an important requirement for traffic simulators.

There exist several specific simulators for urban traffic. AnyLogic[3] is a proprietary software that can be used for that purpose. OpenTrafficSim,[4] in turn, is an open source project for traffic simulation. SimWalk[5] enables urban planners to predict the impact of pedestrians on the road design. Traffic3D is an agent-based traffic simulator that used Unity 3D games engine to animate the simulation (Garg et al., 2022). An entire book was published about the topic some years ago (Barceló, 2010), but recent studies have also still been published about the topic, including results to predict the impact of autonomous driving in modern traffic (Chao et al., 2020; Lu et al., 2020). Chapter 7 discusses mobility in smart cities in a broader perspective.

Analogously, other parts of the city can demand other types of simulators to support their planning. For instance, simulation models have been reported in the literature for smart parking (Delécolle et al., 2020), for the design of energy distribution networks (Lima et al., 2021), smart buildings (Graciano Neto et al., 2020a), Emergency and Crisis Response System (Graciano Neto, 2018), smart home (Bulcão-Neto et al., 2023), and smart lighting (Teixeira et al., 2019). Once each part of a smart city can demand different types of simulation models (and potentially different formalisms), sometimes it can be necessary to use an approach called co-simulation (Gomes et al.,

2018), that is, when engineers need to coexist different simulation models but communicate them to exchange data. For instance, to assure a flood detection in part of a city by an Emergency and Crisis Response System, it would be important to co-simulate the smart traffic control models with the Emergency and Crisis Response System so that a flood detection could trigger changes in traffic.

The following section will go deeper to show how a specific simulation formalism (DEVS) can be used to specify simulation models of parts of a smart city. Further discussions will also be presented.

4.4   HOW TO SPECIFY A SIMULATION FOR A SMART CITY?

DEVS is one of the most used (and powerful) simulation formalisms in the world (de França and Travassos, 2013). In this section, we show how to specify some parts of a smart city using DEVS.

DEVS is a modeling formalism based on atomic and coupled models. Atomic models represent individual entities (for instance, systems), while coupled models represent a combination of atomic models. Coupled models are expressed as a system entity structure (SES), that is, a formal structure governed by a small number of axioms that expresses how atomic models communicate between themselves (Zeigler et al., 2013). DEVS also has a dialect, called DEVS Natural Language (DEVSNL), that enables programming atomic and coupled models in a human-like format in tools such as MS4 Me. Atomic models comprise the following elements: (i) ports, for purposes of simulating input and output; (ii) a state machine that materializes the logic of each constituent being represented; (iii) functions that can be used to process data; (iv) data types; and (v) events (Graciano Neto, 2016).

We use a motivational example of a flood monitoring SoS (FMSoS) (Hughes et al., 2011; Degrossi et al., 2013), which is a recurrent part of smart cities. The first case was an FMSoS in the city of São Carlos, Brazil. That FMSoS refers to a set of independent systems with no central authority to monitor rivers that cross urban areas (where floods can occur and pose great danger in rainy seasons, potentially damaging property, threatening lives, and spreading diseases). It notifies possible emergency situations to residents, business owners, pedestrians, drivers near the flooding area, and governmental entities and emergency systems. Moreover, FMSoS is intended to be part of a larger system composed of wireless river sensors, tele-communication gateways, unmanned aerial vehicles (UAVs), vehicular ad hoc networks (VANETs), meteorological centers, fire and rescue services, hospital centers, police departments, short message service centers, and social networks. Such a system involves the National Center for Natural Disaster Monitoring, which monitors 1,000 cities with 4,700 sensors, including 300 hydrological sensors and 4,400 rain gauges. This FMSoS is deployed over the river; in particular, its sensors are spread on the riverbank's edges, and data is transmitted to gateways. Drones fly over the river and communicate flood threat alerts. People walking close to the river can also communicate the water level increases through mobile applications (Graciano Neto et al., 2022).

FMSoS monitors occurrences of floods in an urban area. Figure 4.3 illustrates an FMSoS in an urban area. As it can be seen in Figure 4.3, a river crosses the city.

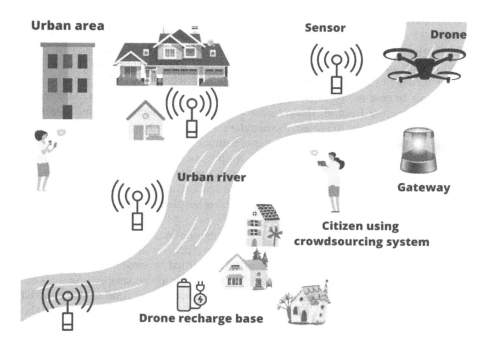

FIGURE 4.3   Illustration of an FMSoS monitoring a river in an urban area.

When the rains are intense, floods frequently occur, causing losses, damage, and imminent danger for the population. FMSoS is composed by five different types of constituents (Graciano Neto et al., 2017):

- Smart sensors, which are fixed embedded systems monitoring flood occurrences in urban areas, located on river edges.
- Gateways, which gather data from constituents and share them with other systems.
- Crowdsourcing systems, which are mobile applications used by citizens for real-time communication of water level rising.
- Drones, which are UAVs also concerned to complement sensors observations by monitoring the river water level while they fly over it, sending pictures if some change in the water level occurs.
- Drone bases, which are fixed bases from where drones depart, and for where they come back to recharge battery, and transmit their data.

FMSoS is concerned with one specific mission: *emitting flood alerts to public authorities that can draw strategies to protect the population*. It consists of a collaborative SoS, with no central authority that orchestrates the constituents functionalities to accomplish missions. Data is gathered in gateways, analyzed according to flood risk, and a status (alert or no alert) is transmitted to public authorities. Sensors are spread on the river's edges with a regular distance among them, and mediators exist between every pair of sensors in a pre-established distance between them. Data collected by sensors is transmitted until reaching the gateway. Besides, drones fly on the river and return to their bases to recharge and eventually communicate with gateways to alert about a flood threat. In parallel, people who walk close to the river can also contribute by communicating that water level is increasing if they perceive this happening. In case of flood, gateways emit alarms for public authorities. Authorities cross data coming from all the constituents to draw a conclusion of an imminent flood, taking decisions to protect the population (Graciano Neto et al., 2017).

For modeling purposes, each of the constituents shown above are modeled as individual atomic models in DEVS. It is important to remark that, besides those constituents, FMSoS also has *mediators*, that is, architectural elements that establish communication between two or more constituents (Garcés et al., 2019). Concrete examples of mediators can be routers, hubs, and other elements that can forward data or reinforce signals.

Since the representation of a smart sensor can be excessively complex for this context, the discussion will be drawn over the specification of mediators; moreover, the specification of the FMSoS will also be simplified to sensors, mediators, and gateway. Figure 4.4 shows the FMSoS under an architecture perspective. Sensors are spread on the river's edges with a regular distance among them, and mediators exist between every pair of sensors that has a distance of 50 meters between them. The data collected by sensors are collected and transmitted until reaching the gateway. In case of a flood, the gateway emits an alarm for the public authorities.

Herein, a mediator only receives the data collected by the sensors and forwards them to the next smart sensor so that it reaches the gateway at some moment. Figure 4.5 shows a simplified code of a mediator type named "transmitter" described in DEVSNL. As it can be observed, the language is quite simple so that it is possible to understand what it is specified. Data types are defined on lines 1–23. This is necessary because the transmitter needs to be aware of the types of data that it can receive and forward. The data being transferred by a smart sensor contain a

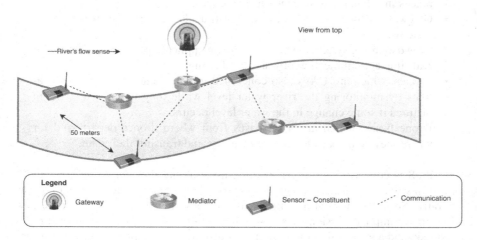

FIGURE 4.4   Illustration of a simplified FMSoS.

```
1    A Distance has a value!
2    the range of Distance's value is Integer!
3    use distance with type Distance!
4
5    A Abscissa has a value!
6    the range of Abscissa's value is Integer!
7    use abscissa with type Abscissa!
8    A Ordinate has a value!
9    the range of Ordinate's value is Integer!
10   use ordinate with type Ordinate!
11   Coordinate has x and y!
12   the range of Coordinate's x is Abscissa!
13   the range of Coordinate's y is Ordinate!
14   use coordinate with type Coordinate!
15
16   A Depth has a value!
17   the range of Depth's value is Integer!
18   use depth with type Depth!
19   Measure has coordinate and depth!
20   the range of Measure's coordinate is Coordinate!
21   the range of Measure's depth is Depth!
22   use measure with type Measure!
23
24   accepts input on FromCoordinate with type
         Coordinate!
25   accepts input on ToCoordinate with type Coordinate
         !
26   accepts input on FromSensors with type Measure!
27   generates output on Measure with type Measure!
28
29   to start hold in s0 for time 1!
30   hold in s0 for time 1!
31   from s0 go to s1! //Unobservable
32   passivate in s1!

33   when in s1 and receive Coordinate go to s2!
34   passivate in s2!
35   when in s2 and receive Coordinate go to s3!
36   passivate in s3!
37   when in s3 and receive Measure go to s4!
38   hold in s4 for time 1!
39   after s4 output Measure!
40   from s4 go to s5!
41   hold in s5 for time 1!
42   from s5 go to s3! //Unobservable
```

FIGURE 4.5 An atomic model for a mediator generated in DEVS.

```
1   From the top perspective,
        FloodMonitoringSosArchitecture is made of
        Sensor1, Sensor2, Sensor3, Sensor4, Gateway,
        Mediator1, Mediator2, Mediator3, and
        Mediator4!
2   From the top perspective, Sensor1 sends Measure to
        Mediator1!
3   From the top perspective, Mediator1 sends Measure
        to Sensor2!
4   From the top perspective, Sensor2 sends Measure to
        Mediator2!
5   From the top perspective, Mediator2 sends Measure
        to Gateway!
6   From the top perspective, Sensor3 sends Measure to
        Mediator3!
7   From the top perspective, Mediator3 sends Measure
        to Sensor4!
8   From the top perspective, Sensor4 sends Measure to
        Mediator4!
9   From the top perspective, Mediator4 sends Measure
        to Gateway!
```

FIGURE 4.6   Coupled model for FMSoS generated in DEVS.

coordinate (composed of abscissa and ordinate) and a depth, which correspond to the level of water that the river exhibits at the moment the sensor measures it. A measure is an information composed of the coordinate from where it was collected and the water depth. Distance (lines 1–3) is only for processing purposes. Lines 24–27 define the input and output ports and lines 29–42 specify the state machine that guides the mediator's operation. In short, it receives the coordinate and the measure, and forwards that information.

Figure 4.6 shows a coupled model of an FMSoS in DEVS, that is, the specification of how the constituents communicate so that the emergent behaviors can manifest themselves. The FMSoS architecture is composed of the sensors, gateway, and mediators (line 1). Sensors transmit their data to the closest transmitter (lines 2, 4, 6, and 8). Then, transmitters forward these data in lines 3, 5, 7, and 9 to a nearby sensor or gateway. Since Sensor2 and Sensor4 send data to Mediator2 and Mediator4 (lines 4 and 6), respectively, the gateway is already reached (lines 5 and 9).

4.5 CHALLENGES IN SIMULATION FOR SMART CITIES

After presenting an overview and advances in the simulation for smart cities, this section aims to name the pain, exposing the challenges that remain in regard to the adoption of simulation in the life cycle of smart cities, analogously to what is done in Graciano Neto et al. (2022).

- Quality assessment via simulation. Quality is about the degree to which a system is close to what is expected from the client. In that sense, it is common to think about quality attributes, which are properties of a system that are not functionalities themselves (such as withdraw money in a ATM), but it is related to them, for instance, how correct the withdraw was and how much time the withdraw took (how fast it was). Since they are qualities, we often use adjectives to qualify the systems, such as how fast it is (whose corresponding quality attribute is named performance), how available it is (whose corresponding quality attribute is named availability), and so on. M&S have been adopted to evaluate quality attributes of systems in smart cities context, such as functional suitability (Graciano Neto et al., 2020a), availability (Delécolle et al., 2020) and reconfigurability (Manzano et al., 2020). Each quality attribute requires specific strategies of representation and analysis in the simulation model, such as the creation of artificial entities to support the analysis, observation of the simulation dynamics, or analysis of the output log. Hence, dealing with a larger set of quality attributes being prioritized in a large-scale context as a smart city can be significantly onerous and require research on how to precisely represent and analyze each relevant quality attribute for that system and its respective domain. Then, further research is needed to generate and catalog techniques to precisely represent each relevant quality attribute that each part of a smart city should satisfy, besides investigating how to co-exist all these underpinning specification details in the final simulation model.
- Trust and reliability in the simulation model and its results. One recurrent question for those who use M&S (or any type of model) is "How do I know that the model is precise enough so that I can trust the results being delivered?" This has been a philosophical question since engineers started to invest in models. An entire book addresses simulation models validation (Beisbart and Saam, 2019). We are aware that the absence of modeling activities in systems development often brings negative impacts on it. Models are a type of abstraction, that is, details necessarily and intentionally are suppressed to represent only the relevant aspects of that system of interest. This difficulty is the same regarding other types of models. At first glance, a simulation model should be precise enough if it covers all (or the majority of) the relevant aspects needed to draw conclusions about that system. Hence, the first advice is to involve a stakeholder highly skilled in modeling in the development team. With his/her experience, she/he can help on the level of details the simulation model should exhibit to be minimally reliable about its results. As a classical and already well-succeeded agile practice, a peer review would also be welcome. Nevertheless, in M&S community, they have adopted a solution called multiresolution modeling (MRM) (Zeigler et al., 2018), that is, they prescribe the adoption of multiple and complementary models besides a concept of multiple instances sizes to analyze the problem. They advocate that MRM is essential for exploratory analysis of the design spaces of complex and adaptive systems as smart ecosystems because it is neither cognitively nor computationally possible to

keep track of all relevant variables and causal relationships (Zeigler et al., 2018). Their concept is close to the induction principle from computer science: they test for a small instance and some larger instances and assume, by induction, that larger frames are also feasible, varying the resolution with which the simulation model is observed and modeled. In practice, this would correspond to modeling a smart city with two constituent systems, another one with three constituents, another with ten, another with a hundred, and then analytically predict how it could behave for larger instances, which could help on the assurance of the reliability of the simulation model and how close it is to the real-world counterpart being represented. A last desirable characteristic to assure simulation model trustworthiness leads to the next challenge.

- Status quo of simulation in the software loop and the presence of professionals of that specific domain. Currently, the basic process that to incorporate simulation models in P&D projects can be: step 1 – understanding the problem, step 2 – a mapping between problem and space solution, step 3 – a system specification, step 4 – design of the simulation model by mapping the system specification into the simulation formalism basic concepts, step 5 – simulation execution, observation, and logging, step 6 – analysis of results, and step 7 – smart city development itself. A real case that can be taken into account is the M&S of the Brazilian space system (Graciano Neto et al., 2018c). The research team luckily could count on an expert in space systems, that is, a Ph.D. student who was in an internship in their laboratory at that time. He fed them with all the needed expertise that allowed them to precisely model that system. Moreover, he also acted as an evaluator for the quality and reliability of their simulation model. Then, the presence of an expert in the domain for which the simulation is being specified is essential for assuring the trustworthiness of the model being developed (in the context of a software system development or not). Another possible solution could be the joint adoption of domain-specific simulators.
- Scalability and simulation costs. Another issue about M&S regards its cost from a diversity of perspectives, including the man-hour cost to specify the model and the execution costs. The former issue is another recurrent concern when we deal with modeling at large. Modeling has a cost, but not modeling can be even more expensive. Simulation modeling still requires the presence of a professional with simulation expertise, who can be roughly expensive and/or rare due to the first challenge we raised (lack of training among the developers). However, for critical domains where failures can lead to significant losses, the cost involved in hiring a simulation expert is undoubtedly lower than the consequences of an eventual failure due to a non-exhaustive design-time evaluation or testing. The latter issue regards the needed infrastructure to support a large-scale simulation execution. Some simulations with only a couple of hundreds of constituents (around 300) can take more than 30 hours to simulate using powerful processors. For simulating smart cities, if we consider the population of New York city or São Paulo and the number of cell phones, autonomous

cars, smart buildings, houses, and hospitals, with different granularities that could potentially compose that ecosystem, the simulation could reach about 50 million constituents. Currently, we do not have computational processing power or simulation techniques to tame such complexity. As stated before, techniques such as the MRM should be adopted to avoid such a high cost, framing different subparts of that larger system and abstracting parts to represent them interacting as black boxes. Co-simulation, as mentioned in Section 4.3, could also be used for that purpose. However, co-simulation itself is also a significant challenge, and we discuss it as follows.

SIMULATE OR NOT SIMULATE: THAT IS THE QUESTION!!

After walking on the challenges, engineers still face to simulate any system, newcomers could ask themselves: but given all these challenges, should I use simulation in my projects? Well, simulation is certainly a mature technology available for all engineers. In general, it is strongly recommended for critical domains, that is, those in which failures and/or malfunction could cause injuries, threats, financial or environmental losses, or other types of losses. Smart cities are certainly critical domains. Failures in traffic systems could lead to collisions and several types of losses, including human, monetary, public, and private. Problems in software, hardware, and other parts controlling autonomous cars, water management systems, or flood monitoring systems could certainly bring serious problems for urban living.

Then, at least for smart cities, when we ask "simulate or not simulate," it is recommended to simulate. Modeling a system is a necessary step to understand its behavior and make predictions, and not doing it can result in negative impacts. To reinforce reliability in the conceived models, the level of details in a simulation model should be determined by involving a skilled modeling stakeholder and through peer review. To ensure the trustworthiness of the simulation model, the MRM approach can be adopted, which involves testing multiple instances of varying sizes to predict the behavior of larger systems. The cost of simulation, including the man-hour cost and the cost of computational processing power, can be high, but it is potentially still lower than the cost of a potential failure due to a non-exhaustive design-time evaluation or testing.

And you? Which failures you envision a smart city could cause due to a malfunction or neglecting engineer plan that does not take simulation into account? Tell us here: https://forms.gle/2AY9R3pu5bL9YpaB9

- Co-simulation. Large companies often have their own simulators or use general-purpose ones, such as MATLAB/Simulink, to specify simulation models and perform their analyses. However, those companies have required researchers to extrapolate the simulation of a single type of system (such as a mechanical or electrical system) toward holistically integrating them into other artifacts and obtain a cohesive simulation that includes several aspects of several domains, such as the combination of simulations of the mechanical, electrical, hydraulic, aerodynamic, hardware, software, people, and environmental parts of the same system (as an aircraft), and how these parts interplay, that is, integrating various simulators to perform together as a composite simulation, which has been termed as co-simulation. As stated by Zeigler et al. (2018), this task involves weaving the time series behavior and data exchanges accurately since a failure on it could yield inaccurate simulation results. Equipping companies with a holistic solution that covers all these aspects is certainly still a challenge. Digital twins, as discussed in Chapter 5, are perhaps an embryonic solution in that direction, but a significant research effort is still required.

- Knowledge absence on simulation. Despite being considered a *de facto* technology for systems engineering, in some other engineering fields, M&S are still not incorporated in their scholar curricula and professional life. A study investigated how much 58 software engineering professionals were prepared to solve daily problems using simulation models (Lebtag et al., 2021). From a set of 58 participants, more than 50% (precisely 30 participants) did not have any contact with simulation models and other types of ExM before. In addition, several participants (even the experienced ones) suggested improvements and opportunities could make ExM technology closer to their current practice, making its adoption easier. The authors concluded that (i) professionals envision several advantages and opportunities for applying ExM in software engineering such as architectural assessment and documentation, (ii) the research field should provide an effective cost-benefit infrastructure for software development before its larger adoption, and (iii) any successful strategy for adoption should be integrated in the software development process smoothly, reducing cognitive stress of learning new technology.

4.6   FINAL REMARKS

In this chapter, we discussed how simulation models can be introduced in the smart cities development life cycle and support engineers to plan their structure and behaviors, besides predicting important properties. Simulation can help in assuring the quality of the systems involved in a smart city. The evaluation of the smart cities architectures could contribute to the quality assurance. Simulation has already proved its value and has provided various benefits for traditional systems evaluation. It has been experimented with and adopted by the community of smart cities, but many developments are still needed. In that sense, the next chapter discusses a correlated topic: digital twins, and how these innovative techniques can also help in the smart cities engineering.

NOTES

1 https://www.mathworks.com/products/simulink.html
2 http://www.netlogoweb.org/

3 https://www.anylogic.com/road-traffic/
4 https://opentrafficsim.org/
5 https://www.simwalk.com/modules/simwalk_roadtraffic.html

REFERENCES

Banks, J. (1999). Introduction to simulation. In Proceedings of the 31st Winter Simulation Conference: Simulation—a Bridge to the Future – Volume 1, ser. WSC '99 (pp 7–13). ACM, New York, NY, http://doi.acm.org/10.1145/324138.324142

Barceló, J. (ed). (2010). Fundamentals of Traffic Simulation (Vol. 145, p. 439). New York, NY: Springer.

Beisbart, C., & Saam, N. J. (2019). Computer Simulation Validation. Springer, New York, NY, Berlin Heidelberg.

Bulcão-Neto, R., Teixeira, P., Lebtag, B., Graciano-Neto, V., Macedo, A., & Zeigler, B. (2023). Simulation of IoT-oriented fall detection systems architectures for in-home patients. IEEE Latin America Transactions, 21(1), 16–26. https://latamt.ieeer9.org/index.php/transactions/article/view/6863

Chackravarthy, S., Schmitt, S., & Yang, L. (2018). Intelligent crime anomaly detection in smart cities using deep learning. In 2018 IEEE 4th International Conference on Collaboration and Internet Computing (CIC) (pp. 399–404). IEEE, Philadelphia, PA.

Chao, Q., Bi, H., Li, W., Mao, T., Wang, Z., Lin, M. C., & Deng, Z. (2020, February). A survey on visual traffic simulation: Models, evaluations, and applications in autonomous driving. Computer Graphics Forum, 39(1), 287–308.

de França, B. B. N., Travassos, G. H. (2013). Are we prepared for simulation based studies in software engineering yet? CLEI Electronic Journal, 16(1), 1–25.

de França, B. B. N., Travassos, G. H. (2016). Experimentation with dynamic simulation models in software engineering: Planning and reporting guidelines. Empirical Software Engineering, 21(3), pp. 1302–1345.

Degrossi, L. C., Do Amaral, G. G., De Vasconcelos, E. S., de Albuquerque, J. P., & Ueyama, J. (2013, May). Using Wireless Sensor Networks in the Sensor Web for Flood Monitoring in Brazil. In ISCRAM.

Delécolle, A., Lima, R. S., Graciano Neto, V. V., & Buisson, J. (2020). Architectural strategy to enhance the availability quality attribute in system-of-systems architectures: A case study. In Proceedings of the System of Systems Engineering Conference 2020 (pp. 93–98), Budapest, Hungary.

Dobrica, L., Niemele, E. (2002, July). A survey on software architecture analysis methods. IEEE Transactions on Software Engineering, 28(7), 638–653.

France, R., & Rumpe, B. (2007). Model-driven development of complex software: A research roadmap. In 2007 Future of Software Engineering, FOSE '07 (pp. 37–54). IEEE Computer Society, Washington, DC, USA.

Garcés, L., Oquendo, F., & Nakagawa, E. Y. (2019). Software mediators as first-class entities of systems-of-systems software architectures. Journal of the Brazilian Computer Society, 25(1), 1–23.

Garg, D., Chli, M., & Vogiatzis, G. (2022). Fully-autonomous, vision-based traffic signal control: From Simulation to Reality. In Proceedings of the 21st International Conference on Autonomous Agents and Multi-agent Systems (AAMAS 2022). In Proceedings of the International Joint Conference on Autonomous Agents and Multiagent Systems, ACM, AAMAS. Auckland, New Zelaand.

Giannoutakis, K. M., Spanopoulos-Karalexidis, M., Filelis Papadopoulos, C. K., & Tzovaras D. (2020). Next generation cloud architectures. In: Lynn T., Mooney J., Lee B., Endo P. (eds) The Cloud-to-Thing Continuum. Palgrave Studies in Digital Business & Enabling Technologies. Palgrave Macmillan, Cham. https://doi.org/10.1007/978-3-030-41110-7_2

Gomes, C., Thule, C., Broman, D., Larsen, P. G., & Vangheluwe, H. (2018). Co-simulation: A survey. ACM Computing Surveys (CSUR), 51(3), 1–33.

Graciano Neto, V. V. (2016). Validating Emergent Behaviours in Systems-of-Systems through Model Transformations. In SRC@ MoDELS (pp. 1–6). ACM, Saint-Malo, France.

Graciano Neto V. V., Santos R. P., Viana D., & Araujo R. (2020b). Towards a conceptual model to understand software ecosystems emerging from systems-of-information systems. In: Santos R., Maciel C., Viterbo J. (eds) Software Ecosystems, Sustainability and Human Values in the Social Web. WAIHCWS 2017, WAIHCWS 2018. Communications in Computer and Information Science, vol 1081. Springer, Cham. https://doi.org/10.1007/978-3-030-46130-0_1

Graciano Neto, V. V. (2018). A simulation-driven model-based approach for designing software-intensive systems-of-systems architectures. (Une approche dirigée par les simulations à base de modèles pour concevoir les architectures de systèmes-des-systèmes à logiciel prépondérant). PhD Thesis. University of Southern Brittany, Vannes, Morbihan, France.

Graciano Neto, V. V., Barros Paes, C. E., Garcés, L., Guessi, M., Manzano, W., Oquendo, F., & Nakagawa, E. Y. (2017). Stimuli-SoS: A model-based approach to derive stimuli generators for simulations of systems-of-systems software architectures. Journal of the Brazilian Computer Society, 23(1), 1–22.

Graciano Neto, V. V., Garcés, L., Guessi, M., Paes, C., Manzano, W., Oquendo, F., & Nakagawa, E. (2018b). ASAS: An approach to support simulation of smart systems. In Proceedings of the 51st Hawaii International Conference on System Sciences (HICSS) (pp. 5777–5786). Waikoloa Village, HI.

Graciano Neto, V. V., Manzano, W., Antonino, P. O., & Nakagawa, E. Y. (2022). Foundations and research agenda for simulation of smart ecosystems architectures. In European Conference on Software Architecture (pp. 333–352). Springer, Cham.

Graciano Neto, V. V., Manzano, W., Kassab, M., & Nakagawa, E. Y. (2018a). Model-based engineering & simulation of software-intensive systems-of-systems: experience report and lessons learned. In Proceedings of the European Conference on Software Architecture (Companion) (27, 1–27:7). Madrid, Spain.

Graciano Neto, V. V., Manzano, W., Rohling, A. J., Ferreira, M. G. V., Volpato, T., & Nakagawa, E. Y. (2018c). Externalizing patterns for simulations in software engineering of systems-of-systems. In Proceedings of the 33rd Annual ACM Symposium on Applied Computing (pp. 1687–1694). Pau, France.

Graciano Neto, V. V., Teles, R. M., Ivamoto, M., Mello, L. H., & de Carvalho, C. L. (2010). Um sistema de apoio à decisão baseado em agentes para tratamento de ocorrências no setor elétrico (An agent-based decision support system for handling incidents in the electricity sector). Revista de Informática Teórica e Aplicada, 17(2), 139–153. In Portuguese.

Graciano Neto, V. V., Horita, F., Santos, R. P., Viana, D., Kassab, M., Manzano, W., & Nakagawa, E. Y. (2020a). S.O.B (Save our budget) – a simulation-based method for prediction of acquisition costs of constituents of a system-of-systems. iSys – Revista Brasileira de Sistemas de Informação, 13, 6–35.

Gray, J., & Rumpe, B. (2016). Models in simulation. Software and Systems Modeling, 15(3), 605–607.

Hanumantha Rao, B., Arun, S. L., & Selvan, M. P. (2019). An electric power trading framework for smart residential community in smart cities. IET Smart Cities, 1(2), 40–51.

Hojaji, F., Mayerhofer, T., Zamani, B., Hamou-Lhadj, A., & Bousse, E. (2019). Model execution tracing: A systematic mapping study. Software and Systems Modeling, 18(6), 3461–3485.

Hughes, D., Ueyama, J., Mendiondo, E., Matthys, N., Horré, W., Michiels, S., ... & Guan, S. U. (2011). A middleware platform to support river monitoring using wireless sensor networks. Journal of the Brazilian Computer Society, 17(2), 85–102.

Ksiksi, A., Al Shehhi, S., & Ramzan, R. (2015, December). Intelligent traffic alert system for smart cities. In 2015 IEEE International Conference on Smart City/SocialCom/ SustainCom (SmartCity) (pp. 165–169). IEEE, Chengdu, China.

Lebtag, B. G., Teixeira, P. G., dos Santos, R. P., Viana, D., & Neto, V. V. G. (2021). Strategies to evolve ExM notations extracted from a survey with software engineering professionals perspective. Journal of Software Engineering Research and Development, 9, 17–1.

Lebtag, Bruno G. A.; Teixeira, Paulo Gabriel; Graciano Neto, Valdemar Vicente. A Systematic Mapping on Executable Models for the Architectural Design of Systems-of-Systems. In: Anais do I Workshop em Modelagem e Simulação de Sistemas Intensivos em Software (MSSIS), 4, 2022, Uberlândia/MG. Anais [...]. Porto Alegre: Sociedade Brasileira de Computação, 2022. p. 11–20. https://doi.org/10.5753/mssis.2022.225662

Lima, R., Kassab, M., & Neto, V. (2021). Discussing the availability quality attribute in systems-of-systems architectures based on a simulation experiment. In Brazilian Symposium on Software Engineering (pp. 416–421). Joinville, Brazil.

Lu, Q., Tettamanti, T., Hörcher, D., & Varga, I. (2020). The impact of autonomous vehicles on urban traffic network capacity: An experimental Analysis by microscopic traffic simulation. Transportation Letters, 12(8), 540–549. https://doi.org/10.1080/19427867. 2019.1662561

Maciel, R. S. P., David, J. M. N., Claro, D. B., & Braga, R. (2017). Full Interoperability: Challenges and Opportunities for Future Information Systems. Grand Challenges in Information Systems for the next 10 years (2016-2026), SBC, Porto Alegre, Brazil, pp. 107–118.

Manzano, W., Graciano Neto, V. V., & Nakagawa, E. Y. (2020). Dynamic-SoS: An approach for the simulation of systems-of-systems dynamic architectures. The Computer Journal, 63(5), 709–731.

Ören, T., Mittal, S., & Durak, U. (2019). Modeling and simulation: The essence and increasing importance. In: Niazi M. A. (ed) Modeling and Simulation of Complex Communication Networks, the IET Book Series on Big Data. (Appendix A: A list of over 750 types of simulation, Appendix B: A list of 900 types of models, and Appendix C: A list of 120 types of input). IET, London.

Ören, T., Zeigler, B. P., & Tolk, A. (eds). (2023). Body of Knowledge for Modeling and Simulation: A Handbook by the Society for Modeling and Simulation International. Springer Nature, Chalm.

Ören, T. I. (2011). The many facets of simulation through a collection of about 100 definitions. SCS M&S Magazine, 2(2) (April), 82–92.

Rocha, V., Alves, L., Graciano Neto, V. V., Kassab, M. A Review on the Adoption of Agile Methods in the Technology Development for Smart Cities. In: WORKSHOP BRASILEIRO DE CIDADES INTELIGENTES (WBCI), 2., 2019, Belém. Anais [...]. Porto Alegre: Sociedade Brasileira de Computação, 2019. https://doi.org/10.5753/wbci. 2019.6748

Rothenberg, J., Widman, L. E., Loparo, K. A., & Nielsen, N. R. (1989). The nature of modeling. In Artificial Intelligence, Simulation and Modeling (pp. 75–92), John Wiley & Sons, Hoboken, NJ.

Selic, B. (2008). Personal reflections on automation, programming culture, and model-based software engineering. Automated Software Engineering, 15(3–4), 379–391.

Sharma, A., Singh, P. K., & Kumar, Y. (2020). An integrated fire detection system using IoT and image processing technique for smart cities. Sustainable Cities and Society, 61, 102332.

Teixeira, P. G., Lebtag, B. G. A., de Oliveira, L. W., de Carvalho, S. T., Veiga, E. F., & de Sousa Rocha, C. (2019). Modeling and simulation of a smart street lighting system. In

Anais do I Workshop em Modelagem e Simulação de Sistemas Intensivos em Software (pp. 44–48). SBC.

Van Tendeloo, Y., & Vangheluwe, H. (2015, April). PythonPDEVS: A distributed parallel DEVS simulator. In SpringSim (TMS-DEVS) (pp. 91–98).

Van Tendeloo, Y., Vangheluwe, H., & Franceschini, R. (2019, December). An introduction to modeling and simulation with (Python (P)) DEVS. In 2019 Winter Simulation Conference (WSC) (pp. 1415–1429). IEEE.

Zeigler, B. P., Mittal, S., Traore, M. K. (2018). MBSE with/out simulation: State of the art and way forward. Systems, 6(4), 1–18.

Zeigler, B. P., Sarjoughian, H. S., Duboz, R., & Soulie, J. C. (2013). Guide to Modeling and Simulation of Systems of Systems. Springer, London.

5 Digital Twin for Smart Cities Transformation*

5.1 INTRODUCTION

A smart city can better manage urban problems by analyzing data collected from all over the city and inducing efficient use of its resources (Yang and Kim, 2021). A smart city is usually equipped with IoT devices, as discussed in Chapter 2, that continuously collect data used to enhance governance and administrative issues (Kumar et al., 2021) and improve mobility, security, environment, and living standards (Abella et al., 2017). Data can be generated from several activities, such as traffic and transportation, power generation, utility provisioning, water supply, and waste management.

The combination of such data using technologies, such as artificial intelligence (AI) and machine learning (ML), enables the creation of Digital Twin (DT). DT can provide real-time monitoring of their physical twins, perform digital simulations of their twins, and are updated according to changes in the physical twins (Kaur et al., 2020).

The DT concept has been considered crucial for digital transformation and has received considerable attention from academia and industry (Fuller et al., 2020). It aims to replicate physical objects (e.g., equipment, devices, people, spaces, systems, and processes) from the real-world context into digital objects and provide simulations or predictions of situations for solving real-world problems (Jeong et al., 2022). The combination of data and intelligence that represent the behavior and context of a physical system offers an interface that allows monitoring past and present operations and making predictions about the future (Grieves, 2014). A DT continuously learns and updates itself from multiple data sources, including historical data, to represent a physical object in near real-time (White et al., 2021).

The main goal of this chapter is to present an overview of the DT concept, especially in the context of smart cities, some applications and methods to design DT, and the challenges for the development and use of DT in smart cities.

5.2 CHARACTERIZATION OF DIGITAL TWIN

The DT concept has garnered significant interest and has become widely popular, mainly in applications like smart manufacturing and Industry 4.0 (Mylonas et al., 2021). The number of scientific publications on DT has grown significantly in the last five years (Ghita et al., 2020; Jones et al., 2020).

---

* Portions of this chapter were contributed by Leonardo Vieira Barcelos and Elisa Yumi Nakagawa.

DOI: 10.1201/9781003348542-5

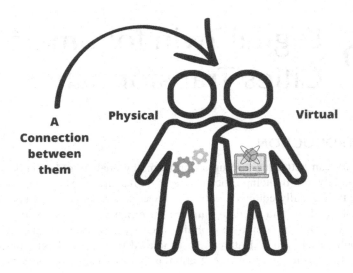

FIGURE 5.1   A ludic illustration of the main elements of DT.

Historically, the first actual DT is associated with the National Aeronautics and Space Administration (NASA) Apollo 13 mission in 1970, which used the DT concept after the explosion of one of the oxygen tanks. The control team in Houston, USA, created ground-level mirrored systems to simulate the conditions of the spacecraft in orbit (Barricelli et al., 2019). "Digital Twin" term was first mentioned in 2003 by Michael Grieves, who introduced the concept in one of his courses about Product Life Cycle Management at the University of Michigan (Grieves, 2014).

Initially, a DT was defined as a digital equivalent to a physical product that requires three main parts to be built (Grieves, 2014): a physical product in the real space, a virtual product in a virtual space, and a connection between them, as picturesquely illustrated in Figure 5.1. With the advancement of research, the concept evolved to meet the needs of different domains. Based on the Semeraro et al. (2021) literature review, DT can be defined as "A set of adaptive models that emulate the behavior of a physical system in a virtual system getting real-time data to update itself along its life cycle." In particular, the DT concept at the city level is denominated as "digital twin city" (or simply DT city). Hence, the physical city can be equated with the corresponding "twin city," allowing them to map each other and interact with each other in both directions (Deren et al., 2021). Nochta et al. (2020) define DT in cities as "digital representations" or "virtual replicas" of cities that can be used in simulation (as explored in Chapter 4) and management environments to develop scenarios in response to policy problems.

There are currently different understandings of DT in the literature, and a consensual definition has not yet been achieved. At the same time, physical entities, virtual models, data, connections, and services are generally considered the core elements of DT (Deren et al., 2021). A DT enables a virtual object to exchange data flow with the physical in both directions (physical-to-virtual and virtual-to-physical

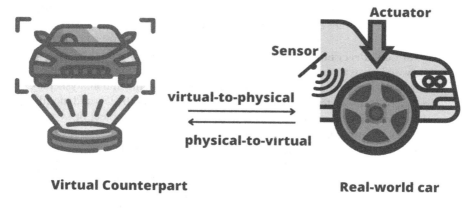

FIGURE 5.2   A model of a general DT for a vehicle.

directions) (Cimino et al., 2019). DT is particularly a confluence of some technologies discussed herein, such as IoT (Chapter 2) and Simulation (Chapter 4).

Figure 5.2 brings a schematic illustration of a DT for a vehicle. As illustrated in Figure 5.2, the data can flow from a digital object to a physical and trigger actions (and changes) in the physical object or instruct actuators to perform an operation (for instance, braking the car autonomously). On the other hand, the data flow from the physical object can influence the digital object in a way that accurately represents the current state and evolution of the physical object, learning and exercising the data, and potentially predicting an imminent problem, alerting the driver or taking decisions.

Kritzinger et al. (2018) propose a classification of DT depending on the interactions (data flow) between the physical and virtual objects: (i) a Digital Model corresponds to a digital representation of a physical object without automated data exchange between the physical and digital objects. Data from physical systems can be used to develop such models, but data exchange is done in a manual way; and (ii) a Digital Shadow corresponds to a digital representation where data flow only happens in one direction, specifically from a physical object to a digital one. A change in the physical object's state leads to a change in the digital object's state, but not vice versa. It includes the Digital Model, data acquisition protocols, and software to simulate the Digital Model.

A Digital Shadow can become a DT when the data flow between the physical and the virtual system becomes bidirectional. Hence, DT can control physical entities without human intervention (Enders and Hoßbach, 2019), map physical systems' attributes, structure, state, performance, function, and behavior to the virtual world in real-time to observe, understand, control, and transform the physical world (Deren et al., 2021), perform simulations as if physical systems are tested in real-world scenarios. Additionally, DT can pause, resume, save, and restore their states to validate different conditions, which would instead be costly and time-consuming, if not impossible, to accomplish with physical systems (Park et al., 2019).

Because there is yet to be a consensus on the definition of DT, different classifications have been proposed in the literature. In terms of operation, DT can be categorized using the following non-necessarily self-excluding taxonomy (Abburu et al., 2020; van der Valk et al., 2020). They are classified in terms of:

- **Data flow direction** as *Read-only*, in which the DT is used exclusively to view and analyze data, but no changes can be made to the physical object or system it represents. This is often used for monitoring and analysis purposes; or *Read-write*, in which the DT allows users to both view and modify data. This may be used to make changes to the physical object or system, such as updating settings or issuing commands;
- **Purpose**, being *Simulative,* when the DT is used to simulate different scenarios or test different approaches. This may be used to optimize performance or identify potential issues before they occur; and/or *Collaborative,* when the infrastructure is shared among multiple users, allowing for collaboration and communication among team members; and
- **Operation mode**, being *Continuous,* with the DT being updated in real-time or near real-time as data is gathered from the physical object or system. This allows for ongoing monitoring and analysis of the physical object or system; or *On-demand*, a mode in which he DT is accessed and used as needed rather than being continuously updated. This may be suitable for less complex or less critical systems.

Another categorization is presented by Ramu et al. (2022), which defined the operation of DT into five levels of sophistication. Each level requires a greater degree of digital maturity and offers increased value to the business, being:

1. **Descriptive twin:** A descriptive twin is a DT used to describe the current state of a physical object or system. It may be used to monitor and analyze data in real-time or near real-time but does not have the capability to make changes to the physical object or system;
2. **Informative twin:** An informative twin is a DT used to provide information about the physical object or system it represents. It may be used to answer questions or provide insights about the physical object or system but does not have the capability to make changes to it;
3. **Predictive twin:** A predictive twin is a DT used to predict future performance or identify potential issues before they occur. It may be used to optimize performance or identify problems in advance, allowing for proactive maintenance or other actions to be taken;
4. **Comprehensive twin:** A comprehensive twin is a highly detailed DT that includes a wide range of data and capabilities. It may be used for various purposes, including simulation, optimization, monitoring, and analysis; and
5. **Autonomous twin:** An autonomous twin is a DT capable of making decisions and taking actions on its own, without human intervention. It may be used to automate processes or control complex systems.

The taxonomies and classifications are abundant and, in general, are based on the DT's ability to represent individual components or entire systems, the level of integration between physical, virtual, and other enterprise systems, the ability of data analysis and simulation to envision new scenarios, optimize performance or predict problems before they occur, the real-time monitoring and the autonomy to make decisions without human intervention.

According to Deren et al. (2021), DT cities have four major characteristics: accurate mapping, virtual-real interaction, software definition, and intelligent feedback.

- **Accurate mapping:** It corresponds to the digital modeling of urban roads, bridges, lamp covers, manhole covers, and buildings, among others, by arranging sensors on the air, ground, underground, and river levels in the physical city to monitor the city's operating status dynamically and provide accurate information;
- **Virtual-real interaction:** It means that all kinds of "traces" that can be observed in the physical city, such as traces of people, vehicles, and logistics, can be searched in the virtual city since they are generated;
- **Software definition:** DT cities have software platforms that simulate the virtual model based on the physical city, representing the behavior of urban people, events, objects, and others; and
- **Intelligent feedback:** DT can trigger early warnings of possible adverse effects, conflicts, and potential danger in the city through planning, design, simulation, etc.

COMMERCIAL DIGITAL TWINS

Certainly, simulation models such as DEVS (as discussed in Chapter 4) can serve as the basis to build DTs in the reality. However, if you want to use to commercial tool to facilitate the process, some of them are available.

In the Body of Knowledge for Modeling and Simulation, Santos and Clua highlight that Microsoft Azure Digital Twins is a powerful commercial tool currently available. The authors emphasize that, in a near future, it is expected that almost all factories, big corporations, and complex scenarios will have their virtual clones, allowing to simulate, monitor, and intervene in any kind of situation or running processes.

Read more at: Santos and Clua (2023). Simulation Games. In Body of Knowledge for Modeling and Simulation: A Handbook by the Society for Modeling and Simulation International (pp. 141–148). Cham: Springer International Publishing.

5.3 APPLICATIONS OF DIGITAL TWIN

DT applications have progressively been consolidated in the last years due to their potential and benefits in different fields, such as manufacturing (Aivaliotis et al., 2019), aerospace (Ye et al., 2020), healthcare (Liu et al., 2019), smart cities (Sta et al., 2021), transportation (Qian et al., 2022), energy (Francisco et al., 2020), construction (Opoku et al., 2021), retail (Augustine, 2020), telecommunication (Hinduja et al.,

2020), and finance (Miskinis, 2019). Henceforth, this section shows how DT concept can be used in several areas of interest that can be used in a smart city from many specialties perspectives.

5.3.1 MANUFACTURING INDUSTRY

In the context of manufacturing industries, the literature often portrays the future as based on DT and cyber-physical systems (CPSs) (also referred to as cyber-physical production systems [CPPSs]). CPSs, with their computing resources, provide flexibility to manufacturing systems, while DT, with their ability to synchronize virtual and physical systems, allow to predict and optimize the behavior of the physical system at each stage of the life cycle (Cimino et al., 2019). At the same time, many manufacturing companies are still equipped with traditional machines, making the implementation of DT difficult (Cimino et al., 2019).

The DT can represent a CPS or system component, contributing to improving time-to-market and MRO (Maintenance, Repair, and Overhaul) costs, predicting possible failures, estimating the remaining useful life of individual components through simulation, and monitoring and controlling the manufacturing process at runtime (Park et al., 2019). Hence, the DT provides information about real-world situations and operational status, making it possible to improve the manufacturing system's intelligence (Tao et al., 2019).

According to Semeraro et al. (2021), DT can be adopted in each phase of the product life cycle and can have different functions. In the design phase, for example, it can help designers configure and validate future scenarios and help the decision-makers interpret market demands and customer preferences. In the manufacturing phase, it can help optimize and evaluate production planning and production process behavior in real-time. In the service phase, DT can monitor the state in real-time for predictive maintenance or predict the remaining life of components or products. Hence, one opportunity that has gained prominence is monitoring manufacturing resources and components to prevent unexpected failures. With the knowledge of the health status of these systems, it is possible to correctly schedule maintenance activities avoiding losses in productivity, increases in costs, and delays in deliveries. These estimates often require historical data from past failures (Aivaliotis et al., 2019). In addition, defective equipment can produce substandard products, causing even more damage.

Here are a few examples of how DT can be used in manufacturing:

- **Design optimization:** DT can optimize the design of manufacturing processes, equipment, and products. By simulating different design scenarios and analyzing the results, manufacturers can identify the most efficient and effective approaches;
- **Process optimization:** DT potentially optimizes the operation of manufacturing processes by analyzing data from sensors and other sources to identify inefficiencies and improve performance;
- **Predictive maintenance:** DT offers prediction abilities when equipment is likely to fail, allowing manufacturers to schedule maintenance in advance and reduce downtime;

- **Quality control:** DT can monitor the quality of products as they are being produced, allowing manufacturers to identify and correct problems before products are shipped to customers;
- **Supply chain optimization:** DT acts on the flow of materials and components through the supply chain, reducing costs and improving efficiency;
- Energy optimization: DT can contribute to a sustainable use of energy in manufacturing processes, reducing costs, and improving sustainability;
- **Customer customization:** DT helps to customize products for individual customers, allowing manufacturers to offer more personalized and differentiated products;
- **Training:** Training processes for employees on new processes or equipment can also be benefited, reducing the need for costly on-site training; and
- **Collaboration:** DT can facilitate the collaboration among teams and departments, improving communication and coordination.

5.3.2 HEALTHCARE

The simulation also plays an important role when using DT in healthcare, especially in medical pathway planning, medical activity prediction, medical resource allocation, etc. In this context, Liu et al. (2019) presented an approach that combines cloud technology with DT to create a framework that helps to monitor, diagnose, and predict aspects of the health of elderly patients. The authors used IoT technology and sensors, such as those for ECG (electrocardiogram), blood pressure, body temperature, and pulse. This approach contributes to supervising the crisis warning of the elderly in healthcare services. Healthcare in smart cities are discussed in detail in Chapter 8.

Here are a few examples of how DT can be used in healthcare:

- **Patient care pathways:** DT can be exploited to simulate and optimize patient care pathways, identifying opportunities to improve efficiency and reduce costs;
- **Clinical decision support:** DT can be utilized to provide clinical decision support to healthcare providers, helping them to make more informed decisions about patient care;
- **Population health management:** DT helps to analyze population health data and identify trends and patterns, helping healthcare providers to identify and address public health issues;
- **Drug development:** DT is a resource to simulate the effects of new drugs on patients, allowing for more informed decisions about drug development and testing;
- **Medical training:** DT is a means to train medical students and healthcare professionals, providing a safe and realistic environment for learning;
- **Surgical planning:** DT assists to plan and simulate surgical procedures, improving the precision and accuracy of surgeries;
- **Medical device design:** DT is a feasible solution to optimize the design of medical devices, improving their functionality and reliability;

- **Telemedicine:** Telemedicine consultations can also rely on DT technology, allowing healthcare providers to diagnose and treat patients remotely; and
- **Clinical trial design:** DT is a potential tool to design and optimize clinical trials, improving their efficiency and effectiveness.

5.3.3 AEROSPACE

In the aerospace domain, DT has been applied to optimize the performance and reliability of space vehicles, predicting and resolving maintenance issues, and making the missions safer for the crew. By reducing the uncertainty of prognosis and maintenance intervals, DT improves the reliability and planning of future missions (Ye et al., 2020). To ensure the safety of crew members, DTs are used to test various possible recovery scenarios in case of emergency. Another benefit of using DT is its use to predict errors, making easier and cheaper the maintenance of spacecraft than scheduled maintenance (Singh et al., 2022).

Here are a few examples of how DT can be used in aerospace (Kim, 2022):

- **Flight simulation:** By means of DT, a flight can be simulated before launching of vehicle, thereby maximizing mission success;
- **Flight continuous mirroring:** DT can mirror various factors and obtain updated conditions of an actual flight, including actual load and temperature. Future scenarios can also be predicted;
- **Diagnosis of vehicle damage:** DT can diagnose damage to the vehicle. If any problem is diagnosed, in-situ repair or mission modifications can be made appropriately, resulting in extended life expectancy of the vehicle or an increased probability of mission success; and
- **Evaluation of the effects of design modifications:** DT can evaluate the effects of modifications of parameters that were not considered in the design phase.

5.3.4 TRANSPORTATION

A transportation system provides various services for traffic management control, including applications such as toll payment, vehicle operation, emergency, vehicle control and safety management, and maintenance and construction management. This system requires real-time data transmission and analysis capabilities to implement time-sensitive services, making it difficult for traditional simulation technologies to complete massive data training in a short time. In this case, DT based on high-performance cloud/edge servers can be adopted to conduct data aggregation and training so that real-time processing and transmission can be achieved (Qian et al., 2022).

Here are a few examples of how DT can be used in transportation:

- **Fleet management:** DT can be linked to an entire fleet and optimize the routing and scheduling of vehicles, improving efficiency and reducing costs;
- **Traffic management:** DT can be used to simulate and optimize traffic flow, reducing congestion and improving safety;
- **Vehicle design:** DT can optimize the design of vehicles, improving performance, and reducing emissions;

- **Infrastructure design:** DT aid in the design of transportation infrastructure, such as roads, bridges, and tunnels;
- **Public transportation:** DT can serve as a means to improve the operation of public transportation systems, improving efficiency and reducing costs;
- **Maintenance:** DT can assist with the prediction when vehicles are likely to need maintenance, allowing for proactive maintenance and reducing downtime;
- **Supply chain optimization:** DT can refine the flow of goods through the transportation system, reducing costs and improving efficiency;
- **Energy optimization:** DT enhances the use of energy in transportation, reducing costs and improving sustainability; and
- **Safety:** DT can identify and mitigate potential safety issues in the transportation system, improving safety for both passengers and employees.

5.3.5 ENERGY

In the field of energy, DT has been applied to assess the energy performance of buildings and make efficiency improvements. Energy is one of the most critical resources for the functioning of a city. Regardless of the city's size, it usually requires huge amounts of energy. Interest in grid energy efficiency and the availability of smart energy metering devices have leveraged the adoption of DT in the energy sector (Mylonas et al., 2021). Francisco et al. (2020) used smart meter electricity data to develop daily energy benchmarks, segmented by strategic periods, to quantify the variance from annual benchmarking strategies and investigate how these metrics can lead to realistic energy management in near real-time. The results showed that the temporally segmented building energy benchmarks are distinct from the overall building benchmark, demonstrating that the general reference of a building masks the periods when a building performs superior or inferior. Hence, temporally segmented energy benchmarks can provide a more specific and accurate measure of building efficiency. The authors discussed the importance of DT-enabled energy management platforms in facilitating energy efficiency prioritization and near real-time decision-making.

Here are a few examples of how DT can be used in the energy industry:

- **Power plant optimization:** DT serves as a means to optimize the operation of power plants, reducing costs and improving efficiency;
- **Renewable energy:** DT can be employed to optimize the operation of renewable energy systems, such as wind farms and solar panels, improving their performance and reliability;
- **Grid management:** The operation of electrical grids can improve the reliability of a smart grid and, as a consequence, reduce outages;
- **Predictive maintenance:** DT can predict when equipment is likely to fail, allowing energy companies to schedule maintenance in advance and reduce downtime;
- **Energy trading:** DT support the trading of energy, reducing costs and improving efficiency;
- **Energy storage:** This essential resource for any electric system can be boosted by DT by refining the design and operation of energy storage systems, such as batteries and pumped hydro storage;

- **Energy efficiency:** The identification and resolution of inefficiencies in the energy system can be aided by DT, which can reduce costs and bring environmental benefits;
- **Carbon capture:** DT can assist in the optimization of carbon capture and storage systems, leading to reduced emissions and improved sustainability; and
- **Exploration and production:** DT can be used to optimize the exploration and production of oil and gas, reducing costs and improving efficiency.

5.3.6 Construction

The construction industry is perceived as one of the least digitized industries and also slow to innovation. Consequently, poor productivity is cited as a vital aspect of failure in the sector. Adopting technologies such as Building Information Modeling (BIM) has provided evidence of change in the construction industry. However, the adoption of BIM has been slow due to the risks and challenges associated with its development. In this context, the DT paradigm has the potential to transform the construction industry, increasing productivity through predictive analytics and other challenges that the construction industry faces. Some of the industry's challenges include low productivity, poor industry image, low predictability, structural fragmentation, lack of Research and Development (R&D), and investment in innovation (Opoku et al., 2021).

Here are a few examples of how DT can be used in construction:

- **Design optimization:** The optimization of building and structure designs can be achieved through the use of DT, resulting in improved efficiency and reduced costs.
- **Construction planning:** DT can be utilized to plan and coordinate construction projects, leading to improved communication and reduced delays.
- **Safety:** The identification and mitigation of potential safety issues on construction sites can be facilitated by the use of DT, ultimately improving safety for workers.
- **Quality control:** The monitoring of construction work quality can be achieved through the use of DT, allowing for early identification and correction of problems.
- **BIM:** DT can serve as a virtual representation of a building or other structure as part of BIM systems, allowing for improved visualization and analysis.
- **Asset management:** The management and maintenance of buildings and other structures can be streamlined through the use of DT, resulting in improved efficiency and reduced costs.
- **Sustainability:** DT can be utilized to optimize the sustainability of buildings and other structures, reducing energy consumption and environmental impact.
- **Collaboration:** The facilitation of collaboration among construction teams and stakeholders can be achieved through the use of DT, improving communication and coordination.

- **Training:** The training of construction workers on new processes or equipment can be accomplished through the use of digital simulations, reducing the need for costly on-site training.

5.3.7 RETAIL

Retailers are shifting their strategies to support consumer engagement and understand their buying behavior, whether the store visit or online buying, but not limited to buying. Customer follow-up is a continuous journey since the customer is the main person in the business. In the current shopping trend, it is evident that attracting customers is not the main factor in sustaining the business but also retaining them by providing the best support (Augustine, 2020). DTs can help retailers improve the customer experience by means use data on shopping behaviors and customer personas (Ramu et al., 2022). For example, retailers can recommend the most suitable fashion clothing products to customers.

Here are a few examples of how DT can be used in retail:

- **Store design:** The optimization of physical store design can be achieved through the use of DT, resulting in an improved customer experience and increased sales;
- **Inventory management:** DT can be utilized to optimize inventory management, reducing costs and improving efficiency;
- **Supply chain optimization:** The optimization of the flow of goods through the supply chain can be facilitated by the use of DT, resulting in reduced costs and improved efficiency;
- **Customer experience:** The analysis of customer behavior and preferences can be achieved through the use of DT, allowing retailers to tailor their offerings and improve the customer experience;
- **Personalization:** The customization of products for individual customers can be facilitated by the use of DT, allowing retailers to offer more personalized and differentiated products;
- **Omnichannel retailing:** The optimization of the integration of online and offline channels can be achieved through the use of DT, improving the customer experience and increasing sales;
- **Marketing:** Marketing campaigns can be benefited by the use of DT, improving the targeting and effectiveness of marketing efforts;
- **Collaboration:** The facilitation of collaboration among teams and departments can be achieved through the use of DT, improving communication and coordination;
- **Training:** The training of employees on new processes or systems can be accomplished through the use of DT, reducing the need for costly on-site training.

5.3.8 TELECOMMUNICATION

The digital transformation of society and industry has led to the emergence of new network applications. Furthermore, the number of connected devices has increased rapidly, making the behavior of modern networks highly dynamic and heterogeneous. As a result, modern communication networks have become very complex and expensive to manage. Thus, DT allows you to design solutions for network optimization, troubleshooting, what-if analysis, or upgrade planning (Almasan et al., 2022).

In the context of tower management in the telecommunications industry is of great importance. Industry companies collect more than 250 variables in real-time, such as network frequency, bandwidths, temperature, wind speed, and more. These data are stored in databases that feed the DT of a tower allowing the performance of network analysis, improvement of network transfer algorithms, and several other use cases. By creating a twin of physical assets and generating simulations, equipment performance and lifecycle increase due to early action recommendations (Hinduja et al., 2020).

Here are a few examples of how DT can be used in telecommunications:

- **Network design:** The design of telecommunications networks can be benefited by DT, improving coverage and reducing costs;
- **Network optimization:** Networks are required to be optimized under many perspectives. DT can assist in the operation of telecommunications networks, improving efficiency and reducing downtime;
- **Predictive maintenance:** DT can help to predict when equipment is likely to fail, allowing telecommunications companies to schedule maintenance in advance and reduce downtime;
- **Customer experience:** DT is a means to analyze customer behavior and preferences, allowing telecommunications companies to tailor their offerings and improve the customer experience;
- **Network security:** Networks are frequently a target for attacks. By linking a DT to the network, managers can identify and mitigate potential security issues in telecommunications networks, improving security and reducing risks;
- **Capacity planning:** DT can also optimize the use of network resources, improving efficiency and reducing costs; and
- **Energy optimization:** Energy is an expensive resource whose production can be environmentally impactful. As such, after the energy is delivered in the network, it is essential to optimize its distribution and use so avoiding waste. DT can be used to optimize the use of energy in telecommunications networks, reducing costs and improving sustainability.

5.3.9 FINANCE

Statistics report that more than 45% of bank customers take care of their needs using digital devices instead of interacting physically. With the trend of bank branches to migrate their banking services from physical interaction to digital, whether through mobile or desktop devices, the importance of ensuring the security of customer funds and the quality of digital service increases (Miskinis, 2019). In this sense, the security of these online interactions can be defended and strengthened through DT. Simulation technology for banking security makes it possible to create virtual cyber-attack scenarios from which AI can learn. With this, a set of actions to counterattack a digital breach can be built, increasing the security of customer funds. As an alternative customer service and relationship management solution, financial institutions, with the support of third-party companies, are collecting data on how customers interact with online services. This data can provide valuable insight into how DT can be used to create simulations that support virtual assistants to help resolve customer issues. The virtual assistants that operate on a question-and-answer basis are not sufficient for the need of a conscientious customer.

Here are a few examples of how DT can be used in finance:

- **Risk management:** DT models can be linked to the finance system so that they can be used to monitor, identify and mitigate potential risks in financial systems, improving stability and reducing costs;
- **Fraud detection:** DT can identify and prevent fraudulent activity, improving security and reducing losses;
- **Trading:** DT can optimize the trading of financial instruments, reducing costs and improving efficiency;
- **Compliance:** DTs have the ability to monitor compliance with financial regulations, reducing risks and improving efficiency;
- **Customer experience:** DT is able to analyze customer behavior and preferences, allowing financial institutions to tailor their offerings and improve the customer experience; and
- **Personalization:** DT can customize financial products and services for individual customers, allowing financial institutions to offer more personalized and differentiated products.

5.3.10 APPLICATIONS BY INDUSTRY LEADERS

The study of Tao et al. (2019) presented the state of the art of DT applications in leading companies.

- Siemens[1] developed DT for the planning, operation, and maintenance of a power system in Finland and improved automation, data utilization, and decision-making. It also developed a DT for a wastewater treatment plant to monitor pipelines in real-time, save energy, and forecast failure trends in advance;
- General Electric (GE)[2] proved that the DT paradigm could change how a wind farm is developed, operated, and maintained, increasing operational efficiency by up to 20% compared to the traditional paradigm without DT. In addition, GE developed hardware and software to design a wind farm DT. GE has also applied DT to track a locomotive's life cycle, including design, configuration, establishment, operation, etc., monitoring the real-time condition of each component and optimizing operations in a timely. In healthcare, GE applied DT to streamline the operation of a hospital in terms of bed planning and work allocation;

- British Petroleum (BP)[3] applied DT to address the challenge of monitoring and maintaining oil/gas facilities in remote areas. BP deployed a DT to improve the reliability of an oil facility in Alaska. Airbus has developed a DT assembly line to monitor the production process and optimize operational efficiency;
- SAP SE[4] argues that DT can prevent disasters by providing prognostics and health management (PHM) services to equipment in harsh environments. The DT can simulate situations and predict their future evolution, improving the equipment's safety. A digital inspection, for example, can be more economical than a physical inspection; and
- IBM[5] has applied DT in automated vehicles to analyze engine speed, oil pressure, and other critical parameters, thus effectively preventing breakdowns and increasing the ability to develop a more efficient engine.

5.3.11 SMART CITIES

In general, the introduction of DT in smart cities starts with city administrators (politicians) and has required the involvement of citizens who can act as data providers and users of DT (Mylonas et al., 2021). The building of a DT city requires technical and data foundations. The technical base refers to relevant technologies, such as IoT, cloud computing, big data, AI, and 5G. At the same time, the data foundation refers to big data continuously generated from various sensors and cameras in different parts of the city and the digital subsystems adopted by municipal management. Through these foundations, a DT city can, for example, collect data on the operating status of infrastructure, the deployment of municipal resources, and the flow of people, logistics, and vehicles and use these data for decision-making (Deren et al., 2021).

Here are a few examples of how DT can be used in smart cities:

- **Traffic flows:** These essential element of the smart cities infrastructure can be linked to DT so that they could address the effect of changes in traffic flows on citizens' mobility, besides many other types of simulations, as discussed in Chapter 4;
- **Impact of rains:** Rains are a source of danger for urban population in many parts of the world. Losses caused by these natural phenomena can be minimized by the use of DT, which can help in the assessment of the potential effects of excessive rainfall on the water level of a river or in certain areas/ buildings, allowing better response planning;
- **Pollution:** Kyoto Protocol has established guidelines to reduce pollution derived from urban environments, particularly those related to carbon-related ones. However, urban environments and population also suffer from the noise pollution due to cars and constructions. Pollution, in a broad sense of the word, can be monitored, by measuring and processing data collected related to the air pollution levels and dispersion of noise pollution inside a city structure; and

- **Urban planning:** DT can be used to simulate proposed measures by officials, allowing them to select those that produce the least negative impact on other aspects and citizen involvement and acceptance.

ARE DT ALONE IN EARTH? THE ANSWER IS NO. THEY HAVE OTHER RELATIVES. THE DIGITAL QUADRUPLETS!!

Have you ever played a video game or used a computer program where you control a virtual version of something in the real world? That's kind of like what a DT is: a virtual model (which is essentially any computer model, including a 3D model) linked to the real physical thing, like a building or a machine.

Now, imagine if you could use that virtual model not just for looking at and understanding the physical thing, but also for experimenting with it in a computer simulation, and even for controlling it in real time. That's the idea behind a Digital Quadruplet (DQ).

Niyonkuru and Wainer (2021) proposed the concept of a DQ. In their study, a DQ includes four parts (as illustrated in Figure 5.3): the DT, which is (i) the physical object monitored by sensors and operated by actuators linked to (ii) a 3D virtual model (in their case); (iii) a simulation model of the system being studied, called the Digital Triplet; and (iv) a physical model of the real system for experiments. All of these are linked by some artifact to make sure all of these things work together consistently and can be reused.

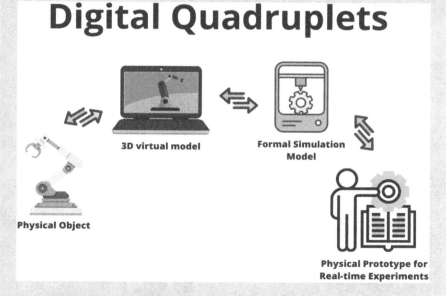

FIGURE 5.3 Digital quadruplets scheme.

To make this happen, the researchers use a Discrete-Event formal model as the "center" for the simulation and real-time control of the system. Think of it like the conductor of an orchestra, making sure everything stays in time and works together.

This method also avoids using an operating system on the computer hardware, which helps keep things running smoothly. By using a DQ, they can experiment and control real physical systems in a safe, virtual environment, which can lead to better designs and improved performance.

Read more at: Niyonkuru and Wainer (2021). A DEVS-based engine for building digital quadruplets. Simulation, 97(7), 485–506.

5.4   DESIGNING DIGITAL TWINS FOR SMART CITIES

White et al. (2021) designed a model composed of six layers of information that can be adopted to design DT for smart cities. In short, the first five layers provide information about the city's terrain, buildings, infrastructure, mobility, and IoT devices, while the last layer uses the data generated in the smart city to perform simulations. As illustrated in Figure 5.4, the layers are:

- **Terrain:** The zeroth layer corresponds to the terrain on which the city is built, containing basic information about the city, such as parts that are

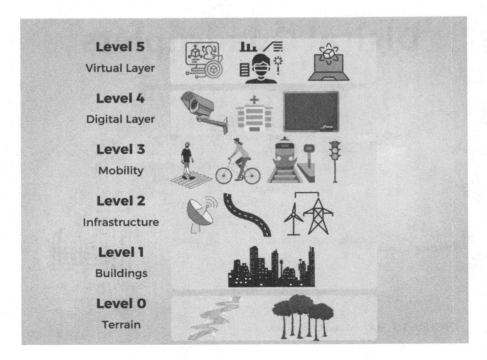

FIGURE 5.4   Smart cities DT structured in layers.

offshore, rivers or canals running through the city, steep gradients or hills, and areas with poor draining soil (e.g., a landslide or flooding during heavy rain). A soil map can be used to incorporate such information into the model;

- **Buildings:** Layer 1 adds buildings into the city model. These buildings feature highly accurate building information modeling (BIM) that can be used as a DT of the building. The 3D building data can also be generated using stereoscopic aerial photography;
- **Infrastructure:** Layer 2 adds the infrastructure surrounding the buildings, such as basic physical and organizational structures and facilities (e.g., roads, power supplies, telecommunications) necessary for the operation of a society or enterprise. This infrastructure data can be obtained from OpenStreetMap,[6] which contains information about energy, public transport, highways, amenities, and telecommunications. Data can also be taken from the 3D mapping process to add gradient information;
- **Mobility:** Layer 3 adds mobility to the infrastructure and buildings layers. It represents the movement of people during their daily routine and the movement of goods. Software applications such as SUMO[7] can be used to simulate urban mobility (Lopez et al., 2018). SUMO simulator supports different transportation modes: walking, bicycles, powered two-wheelers, generic parametrized vehicles, waterways, and railways. SUMO allows being connected to the 3D model in Unity[8] using the Traffic Control Interface (TraCI)[9] and implements and enhances the traffic modes with additional behaviors that the simulator does not model. Figure 5.5 presents an example using SUMO;
- **Digital layer/smart city:** Layer 4 integrates IoT sensors to collect data. This data can be used to monitor and manage traffic and transportation systems, water supply networks, utilities, crime detection, waste management, power plants, hospitals, schools, and libraries, among other services; and

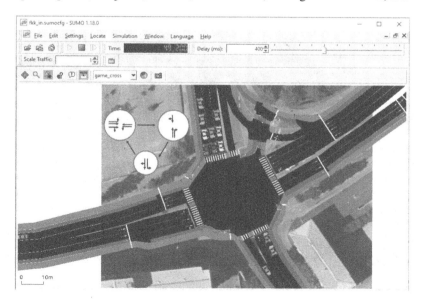

FIGURE 5.5   Application example developed in SUMO.

- **Virtual layer/digital twin:** Layer 5 uses the data produced from the digital layer/smart city to conduct simulations and provides insights regarding the city. This data can also be passed back as information to the model's layers.

The study of Sta et al. (2021) presented a framework to create a DT using open-source software. The essential components of this framework are (Sta et al., 2021): (i) visualization of digital model; (ii) identification of user interface and data management requirements; (iii) user interface set-up and configuration; and (iv) analysis and simulations. The authors explored different tools for city visualization and selected TerriaMap[10] due to its capability to address large areas in 3D and create customizable user interfaces. This tool presents a user-friendly interface that enables performing actions such as datasets being searched, shared, and added to the map and showing dashboards related to management, among other functions. The authors reported they used TerriaJS,[11] an application built based on TerriaMap for user interface set-up and configuration and presented technical instructions for installing and using the application. The analysis and simulations used land surface temperature data from Landsat 8 satellite images. In the example presented, the data were stored on a private server that allows integration with DT. Figure 5.6 shows the application of TerriaJS by the NSW government.

5.5 DISCUSSION AND FUTURE PERSPECTIVES

When properly implemented, DT can benefit several stakeholders in a city (citizens, public administration, asset managers, asset owners, and researchers). Adopting DT in public administration allows administrators to make more informed decisions about city planning and operations, resulting in better management and a more sustainable city (Hamäläinen, 2020). In addition, they can simultaneously process a

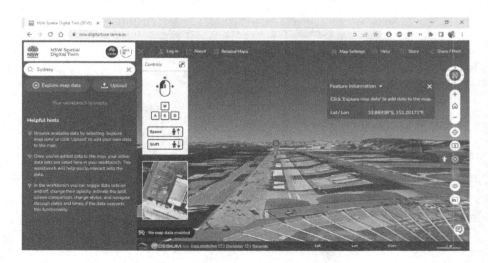

FIGURE 5.6 Application of TerriaJS by the NSW government.

wide variety of datasets from multiple systems and control the city in a single way (Ferré-Bigorra et al., 2022).

Regarding data management, one of the concerns is the need to integrate heterogeneous data due to the large-sized, complexity, and nature of the city data (Shahat et al., 2021). In addition, the data acquisition and processing in real-time require servers with high computational capacity (Ferré-Bigorra et al., 2022). Data quality is critical as inaccurate data directly affects DT outputs and prevents it from providing accurate assistance in making decisions or predicting future scenarios (Nochta et al., 2020). Another issue that offers significant limitations is the lack of standards for interoperability in this domain, resulting in difficulty for urban DT to share data with other organizations (Ferré-Bigorra et al., 2022). Furthermore, Nochta et al. (2020) stated that much data that feeds DT is not easily available due to the dispersed asset and data ownership. They also reported a lack of processes to obtain and manage this data safely and sustainably and to involve multiple actors (public and private organizations and citizens).

Cybersecurity is another concern. If the quality of DT is not assured, it could leak sensitive data. In addition, if attackers gain control of the system, they could seriously affect urban infrastructures, such as traffic management systems, electric power transmission grids, or water supply networks (Lee et al., 2019). Privacy also represents a critical and still open factor for city-scale DT (Mylonas et al., 2021). Data stored in the application layer, for instance, vehicle location and traffic camera data, is confidential, requiring user privacy protection (Qian et al., 2022). In traffic management, there is a risk of abuse of location data produced from vehicles and, consequently, the location of individuals could be tracked (Mylonas et al., 2021).

Other challenges associated with applying DT in smart cities include:

- **Data management:** DT relies on large amounts of data from sensors, monitoring systems, and other sources. Managing this data and ensuring its quality and accuracy can be a challenge;
- **Integration with other systems:** DT must be integrated with other systems and platforms to exchange data and facilitate collaboration. This can be a complex and time-consuming process;
- **Security and privacy:** DT generates and store sensitive data, which must be protected from unauthorized access. Ensuring the security and privacy of this data can be a challenge;
- **Cost:** Implementing DT can be costly, particularly for large-scale projects. This can be a barrier to adoption for some organizations;
- **Complexity:** DT can be complex systems requiring specialized expertise to develop and maintain. This can be a challenge for organizations that do not have the necessary resources or capabilities;
- **Regulation:** DT may be subject to a range of regulatory requirements, depending on the industry and location. Ensuring compliance with these requirements can be a challenge; and
- **Cultural change:** Implementing DT may require significant changes to organizational culture and ways of working. This can be a challenge for some organizations.

Future research should focus on providing solutions to the challenges mentioned earlier. Issues related to standardization, interoperability, and data quality are the most frequent topics in recent research (Mylonas et al., 2021). In addition, communicating the real benefits of DT in terms of informed decisions to the communities involved is another important factor (Mylonas et al., 2021). Issues related to security and privacy are a significant challenge in the implementation and deployment of DT in real-time and require further studies (Ferré-Bigorra et al., 2022).

5.6   FINAL REMARKS

DT has been successfully adopted in the manufacturing industry but is still at an early stage in the smart city domain (Ramu et al., 2022). At the same time, technologies, such as AI, ML, IoT, and cloud computing, are continuously advancing, allowing the development of smart city-based applications in different contexts, such as smart health, transport, and environmental management (Fuller et al., 2020). These applications have become smarter over time and can provide predictive insights about the city's growth and performance (Mohammadi and Taylor, 2017). Moreover, an advantage of DT is integrating city planning and management in a single tool, making it possible to become the city more efficient since such a tool allows knowing the actual state of the city in real-time and being able to actuate autonomously (Ferré-Bigorra et al., 2022).

Recent studies showed the benefits of DT for smart urban management (Barricelli et al., 2019; Guzina et al., 2022). However, the major challenges for the large-scale implementation of the DT concept in urban environments include the availability of data that feeds DT, data quality, data security and privacy, interoperability among various systems and components, and integration of heterogeneous data. Chapter 6 discusses the blockchain technology applied to smart cities.

NOTES

1   https://www.siemens.com/
2   https://www.ge.com/
3   https://www.bp.com/
4   https://www.sap.com/
5   https://www.ibm.com/
6   https://www.openstreetmap.org/
7   https://sumo.dlr.de/
8   https://unity.com/
9   https://sumo.dlr.de/docs/TraCI.html
10   https://github.com/TerriaJS/TerriaMap
11   https://terria.io/

REFERENCES

Abburu S., Berre A. J., Jacoby M., Roman D., Stojanovic L., & Stojanovic N. (2020). Cognitwin–hybrid and cognitive digital twins for the process industry. In 2020 IEEE International Conference on Engineering, Technology and Innovation (ICE/ITMC) (pp. 1–8). IEEE, Cardiff, UK.

Abella A., de Urbina-Criado M. O., & De-Pablos-Heredero C. (2017). A model for the analysis of data-driven innovation and value generation in smart cities' ecosystems. Cities, 64, 47–53.

Aivaliotis P., Georgoulias K., & Chryssolouris G. (2019). The use of digital twin for predictive maintenance in manufacturing. International Journal of Computer Integrated Manufacturing, 32(11), 1067–1080.

Almasan P., Ferriol-Galmés M., Paillisse J., Suárez-Varela J., Perino D., López D., Perales A. A. P., Harvey P., Ciavaglia L., Wong L., Ram V., Xiao S., Shi X., Cheng X., Cabellos-Aparicio A., & Barlet-Ros P. (2022). Digital Twin Network: Opportunities and Challenges (pp. 1–7). https://arxiv.org/pdf/2201.01144v2.pdf, Cornell University.

Augustine P. (2020). The industry use cases for the digital twin idea. Advances in Computers, 117, 79–105.

Barricelli B. R., Casiraghi E., & Fogli D. (2019). A survey on digital twin: Definitions, characteristics, applications, and design implications. IEEE Access, 7, 167653–167671.

Cimino C., Negri E., & Fumagalli L. (2019). Review of digital twin applications in manufacturing. Computers in Industry, 113, 103130.

Dameri R. P., & Rosenthal-Sabroux C. (2014). Smart City and Value Creation (pp. 1–12). Springer International Publishing, Cham.

Deren L., Wenbo Y., & Zhenfeng S. (2021). Smart city based on digital twins. Computational Urban Science, 1(1), 1–11.

Enders M. R., & Hoßbach N. Dimensions of digital twin applications - a literature review.AMCIS 2019 Proceedings. 20, 2019. Available on https://aisel.aisnet.org/amcis2019/org_transformation_is/org_transformation_is/20/

Ferré-Bigorra J., Casals M., & Gangolells M. (2022). The adoption of urban digital twins. Cities, 131, 103905.

Francisco A., Mohammadi N., & Taylor J. E. (2020). Smart city digital twin–enabled energy management: Toward real-time urban building energy benchmarking. Journal of Management in Engineering, 36(2), 1–11.

Fuller A., Fan Z., Day C., & Barlow C. (2020). Digital twin: Enabling technologies, challenges and open research. IEEE Access, 8, 108952–108971.

Ghita M., Siham B., & Hicham M. (2020). Digital twins development architectures and deployment technologies: Moroccan use case. International Journal of Advanced Computer Science and Applications, 11(2), 468–478.

Grieves M. (2014). Digital twin: Manufacturing excellence through virtual factory replication. White Paper, 1, 1–7.

Guzina L., Ferko E., & Bucaioni A. (2022). Investigating digital twin: A systematic mapping study. In Advances in Transdisciplinary Engineering (p. 449–460), v. 21, IOS Press.

Hakak S., Khan W. Z., Gilkar G. A., Imran M., & Guizani N. (2020). Securing smart cities through blockchain technology: Architecture, requirements, and challenges. IEEE Network, 34(1), 8–14.

Hamäläinen M. (2020). Smart city development with digital twin technology. In Bled eConference – Enabling Technology for a Sustainable Society (pp. 291–303). University of Maribor Press, Bled, Slovenia.

Hinduja H., Kekkar S., Chourasia S., & Chakrapani H. B. (2020)Industry 4.0: Digital twin and its industrial applications. International Journal of Science, Engineering and Technology, 8(4), p. 1–7,. Available online: https://www.researchgate.net/publication/343713676_Industry_40_Digital_Twin_and_its_Industrial_Applications (accessed on 19 January 2022).

Jeong D.-Y., Baek M.-S., Lim T.-B., Kim Y.-W., Kim S.-H., Lee Y.-T., Jung W.-S., & Lee I.-B. (2022). Digital twin: Technology evolution stages and implementation layers with technology elements. IEEE Access, 10, 52609–52620.

Jones D., Snider C., Nassehi A., Yon J., & Hicks B. (2020). Characterising the digital twin: A systematic literature review. CIRP Journal of Manufacturing Science and Technology, 29, 36–52.

Kaur M. J., Mishra V. P., & Maheshwari P. (2020). The Convergence of Digital Twin, IoT, and Machine Learning: Transforming Data into Action (pp. 3–17). Springer International Publishing, Chalm, Switzerland.

Kim M. U. (2022). A survey on digital twin in aerospace in the new space era. In 13th International Conference on Information and Communication Technology Convergence (ICTC) (pp. 1735–1737). IEEE Jeju Island, Korea

Kritzinger W., Karner M., Traar G., Henjes J., & Sihn W. (2018). Digital twin in manufacturing: A categorical literature review and classification. IFAC-PapersOnLine, 51(11), 1016–1022.

Kumar P., Kumar R., Srivastava G., Gupta G. P., Tripathi R., Gadekallu T. R., & Xiong N. N. (2021). PPSF: A privacy-preserving and secure framework using blockchain-based machine-learning for IoT-driven smart cities. IEEE Transactions on Network Science and Engineering, 8(3), 2326–2341.

Lee J., Kim J., & Seo J. Cyber attack scenarios on smart city and their ripple effects. In International Conference on Platform Technology and Service (PlatCon), IEEE, 2019, p. 1–5.

Liu Y., Zhang L., Yang Y., Zhou L., Ren L., Wang F., Liu R., Pang Z., & Deen M. J. (2019). A novel cloud-based framework for the elderly healthcare services using digital twin. IEEE Access, 7, 49088–49101.

Lopez P. A., Behrisch M., Bieker-Walz L., Erdmann J., Flötteröd Y.-P., Hilbrich R., Lucken L., Rummel J., Wagner P., & Wießner E. Microscopic traffic simulation using sumo. In IEEE International Conference on Intelligent Transportation Systems, IEEE, 2018, p. 2575–2582.

Miskinis C. (2019). Disrupting the Financial and Banking Services Using Digital Twins. https://www.challenge.org/insights/digital-twin-for-finance/ (accessed on 19 January 2022)

Mohammadi N., & Taylor J. E. (2017) Smart city digital twins. In Symposium Series on Computational Intelligence (SSCI) (pp. 1–5). IEEE, Honolulu, HI.

Mylonas G., Kalogeras A., Kalogeras G., Anagnostopoulos C., Alexakos C., & Muñoz L. (2021). Digital twins from smart manufacturing to smart cities: A survey. IEEE Access, 9, 143222–143249.

Niyonkuru D., & Wainer G. (2021). A DEVS-based engine for building digital quadruplets. Simulation, 97(7), 485–506.

Nochta T., Wan L., Schooling J. M., & Parlikad A. K. (2020). A socio-technical perspective on urban analytics: The case of city-scale digital twins. Journal of Urban Technology, 28(1–2), 263–287.

O'Brien P., Pike A., & Tomaney J. (2019). Governing the 'ungovernable'? Financialisation and the governance of transport infrastructure in the london 'global city-region. Progress in Planning, 132, 100422.

Opoku D.-G. J., Perera S., Osei-Kyei R., & Rashidi M. (2021). Digital twin application in the construction industry: A literature review. Journal of Building Engineering, 40, 1–15.

Park H., Easwaran A., & Andalam S. (2019). Challenges in digital twin development for cyber-physical production systems. In Cyber Physical Systems. Model-Based Design (pp. 28–48). Springer International Publishing, Chalm, Switzerland.

Qian C., Liu X., Ripley C., Qian M., Liang F., & Yu W. (2022) Digital twin—Cyber replica of physical things: Architecture, applications and future research directions. Future Internet, 14(2), 1–25.

Ramu S. P., Boopalan P., Pham Q.-V., Maddikunta P. K. R., Huynh-The T., Alazab M., Nguyen T. T., & Gadekallu T. R. (2022). Federated learning enabled digital twins for smart cities: Concepts, recent advances, and future directions. Sustainable Cities and Society, 79, 1–13.

Santos R. P. D., & Clua E. W. G. (2023). Simulation Games. In Body of Knowledge for Modeling and Simulation: A Handbook by the Society for Modeling and Simulation International (pp. 141–148). Springer International Publishing, Cham.

Semeraro C., Lezoche M., Panetto H., & Dassisti M. (2021). Digital twin paradigm: A systematic literature review. Computers in Industry, 130, 103469.

Shahat E., Hyun C. T., & Yeom C. (2021). City Digital twin potentials: A review and research agenda. Sustainability, 13(6), 3386.

Singh M., Srivastava R., Fuenmayor E., Kuts V., Qiao Y., Murray N., & Devine D. (2022) Applications of digital twin across industries: A review. *Applied Sciences*, 12(11), 1–28.

Sta, Ana R. R., Escoto J. E., Fargas D., Panlilio J., Jerez K., & Sarmiento M. (2021). Development of a digital twin for the monitoring of smart cities using open-source software. International Archives of the Photogrammetry, Remote Sensing and Spatial Information Sciences, 46W6, 281–288.

Tadili J., & Fasly H. (2020). General smart city experts' perceptions of citizen participation: A questionnaire survey. In Innovations in Smart Cities Applications Edition 3 (pp. 3–15). Springer International Publishing, Cham.

Tao F., Zhang H., Liu A., & Nee A. Y. C. (2019). Digital twin in industry: State-of-the-art. Transactions on Industrial Informatics, 4, 2405–2415.2019.

van der Valk H., Hunker J., Rabe M., & Otto B. (2020). Digital twins in simulative applications: a taxonomy. In 2020 Winter Simulation Conference (WSC) (pp. 2695–2706). IEEE, Orlando FL.

White G., Zink A., Codecá L., & Clarke S. (2021). A digital twin smart city for citizen feedback. Cities, 110, 103064.

Yang S., & Kim H. (2021). Urban digital twin applications as a virtual platform of smart city. International Journal of Sustainable Building Technology and Urban Development, 4, 363–379.

Ye Y., Yang Q., Yang F., Huo Y., & Meng S. (2020). Digital twin for the structural health management of reusable spacecraft: A case study. Engineering Fracture Mechanics, 234, 107076.

Zhang X. Q. (2016). The trends, promises and challenges of urbanisation in the world. Habitat International, 54, 241–252.

Zheng C., Yuan J., Zhu L., Zhang Y., & Shao Q. (2020). From digital to sustainable: A scientometric review of smart city literature between 1990 and 2019. Journal of Cleaner Production, 258, 120689.

6 Blockchain for Secure and Transparent Smart City Transactions

6.1 INTRODUCTION

Blockchain technology, the backbone of cryptocurrencies like Bitcoin, is a decentralized, distributed ledger that enables secure, transparent, and tamper-proof transactions. It has the potential to revolutionize a wide range of industries, including smart cities. While blockchain had been created initially to support decentralized financial transactions, currently this disruptive technology has been dramatically applied in various other domains.

The use of blockchain technology in smart cities is a relatively new concept that has gained attention in recent years. The potential benefits of blockchain technology in smart cities, such as improved transparency, facilitated smart contract-based automation of city services and systems, improved communication and coordination among city departments and with citizens, more efficient and secure management of city assets, and the creation of new economic opportunities by enabling the development of new decentralized applications, have led to a growing interest in exploring its use in urban environments. The first known use of blockchain technology in smart cities was in Dubai in 2016, when the government announced plans to use blockchain technology to create a smart city. Dubai's government considers blockchain as an essential layer of trust in a smart city (Kundu, 2019). The goal was to create a fully digital and paperless government, by using blockchain to secure and streamline government services and transactions. With the blockchain, Dubai is expected to save 100 million pages of documents every year, generate 25.1 million hours of economic productivity in savings each year, and enable citizens to save 411 million kilometers of city service-related travel every year (Xie et al., 2019).

The city of Singapore is another prime example of a smart city that has successfully implemented blockchain technology to improve the efficiency of city services and reduce costs. In 2016, the city-state's government launched a program called "Smart Nation" with the goal of using technology to improve the lives of citizens and make the city more efficient (Xie et al., 2019). A key component of the program is the use of blockchain technology to improve the efficiency of government services and reduce costs. One of the most notable examples of the city's use of blockchain technology is its "smart port" project. Singapore's port is one of the busiest in the world, handling over 36 million TEUs (20-foot equivalent units) annually. The government recognized the potential for blockchain technology to improve the efficiency of the port by reducing the need for intermediaries and streamlining the process of tracking and managing shipping containers. To accomplish this, the government developed a blockchain-based platform called "TradeTrust," which is used to securely and transparently track the movement of shipping containers through the port. The platform allows for the automation of many processes, such as the tracking of container movements and the management of customs clearance, which were previously done manually. This has resulted in significant cost savings and increased efficiency.

Another example of Singapore's use of blockchain technology is in the area of digital payments. The government has developed a blockchain-based platform called "Project Ubin" which is used to facilitate digital payments and reduce the need for intermediaries. The platform allows for the creation of digital tokens that can be used to make payments, which can be done quickly and securely without the need for intermediaries.

While these examples illustrate how blockchain technology can be used to improve the efficiency of government services and reduce costs; the implementation of blockchain technology in smart cities is not without challenges. These include scalability, interoperability, regulation, security/privacy, adoption and education, and power consumption.

This chapter discusses how blockchain can be used as a platform in the context of smart cities. This chapter also sheds the light on the open challenges of concern.

6.2   BACKGROUND

Blockchain is the technology behind Bitcoin, Ethereum, and other cryptocurrencies. The theory behind Bitcoin as a peer-to-peer electronic cash system was introduced in a white paper written under the pseudonym "Satoshi Nakamoto" in 2008. A decade later, and despite the uncertainty of the identity of its creator, Bitcoin was rapidly implemented and widely accepted as a prominent online cryptocurrency. This is evidenced by the current estimates of the total value of all Bitcoins in existence which hovers around $586.48 billion (as of July 2023). Many online retailers accept Bitcoin as a means of payment with many mechanisms in existence for exchanging it with a fiat currency and vice versa (Kassab et al., 2019).

The blockchain is the essence of the infrastructure underlying Bitcoin and other cryptocurrencies. Directly speaking, a blockchain constructs a chronological chain of blocks, hence the name. It works as a write-only database in which records can only be added, and all the information is only effectively recorded if there is a consensus among the peers. A peer is a computer (or a pool of computers) managed by a person, entity, or company that is involved in the blockchain and controls the registration of the block, in a manual, semi-automated, or automated manner. Each block: (i) bundles a set of transactions, (ii) is time-stamped, and (iii) is linked to its precedent block (with an exception for the genesis block). Combined with cryptographic hashes, this time-stamped chain of blocks provides an "immutable" record of all transactions in the network, from the original block.

Figure 6.1 brings a conceptual illustration of blockchain architecture. Entities (also referred to as nodes) maintain a blockchain and are connected through a peer-to-peer network without a pre-existing trust. Each node hosts an identical copy of a blockchain (replicated copy of the ledger) creating a decentralized structure.

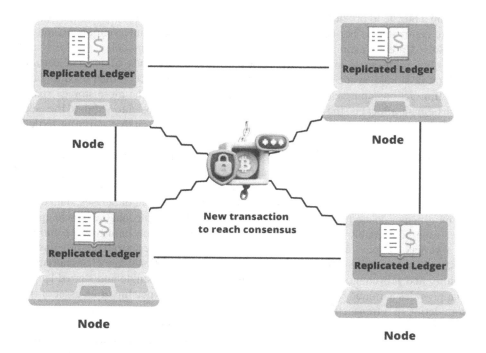

FIGURE 6.1   Blockchain structure.

However, for such a network to be correctly functional, it is crucial to have a protocol by which the nodes can mutually reach a consensus on which transactions are legitimate and to be added to the next valid block in the chain. Some of the utilized schemes for establishing a distributed consensus include Proof of Work (PoW), Proof of Stake (PoS), Proof of Human-Work (PoHW), and Proof of Capacity (PoC).

The first one (PoW) has network participants known as miners, who must solve complex mathematical problems to validate transactions and create new blocks in the blockchain. "Mining" is exactly the computational process in which the miners (powerful machines) spend much electricity and their owners earn money as cryptocurrencies, such as Bitcoin or Ether. However, the related cost and demanded energy for the coin production often do not worth the effort, particularly with the expressive reduction in Bitcoin's market cap between 2021 and 2022. Bitcoin (BTC) is a PoW-based blockchain, and Ethereum (ETH) is transitioning to a PoS.

PoS is a consensus mechanism in which validators (known as stakers) must hold a certain amount of cryptocurrency to validate transactions and create new blocks in the blockchain. The more cryptocurrency a staker holds, the greater their chance of being chosen to validate the next block. PoS is considered to be more energy-efficient than PoW, as it does not require the same level of computational power. Examples of blockchains that use PoS consensus are Cardano (ADA) and Polkadot (DOT).

PoHW involves the completion of a task by a human being to validate transactions and create new blocks in the blockchain. The task may be anything from solving a puzzle to providing an answer to a question. PoHW is designed to be resistant to automated or bot-based attacks, as it requires human intervention. hCaptcha and Solve.Care are examples of blockchains adherent to that model.

In PoC consensus mechanism, network participants must allocate a certain amount of storage space to the blockchain. This storage space is then used to validate transactions and create new blocks in the blockchain. The more storage space a participant allocates, the greater their chance of being chosen to validate the next block. Burstcoin and Chia Network are examples of blockchain that conforms to that model.

A blockchain is a subset of a larger family known as distributed ledger technologies (DLTs). A blockchain is an append-only distributed ledger and all blockchains share five well-defined properties: *decentralization*, *consensus*, *immutability*, *finality*, and *provenance*. In contrast with traditional relational databases, data can't be modified or deleted once recorded. The immutability of the data (i.e., unchanging over time) is arguably the convincing reason to deploy blockchain-based solutions in a variety of socio-economic scenarios. In addition to decentralization, consensus, and immutability, a blockchain network has two additional key characteristics: finality and provenance (Gupta, 2017). Finality refers to a single and shared ledger that provides a unique place to support determining the ownership of an asset or the completion of a transaction. While provenance refers to that participants of the network know where the asset originated and its ownership history.

A blockchain can also use smart contracts, which serve as a set of rules that oversee a business transaction. A smart contract is stored on the blockchain and is executed automatically as part of a transaction. For example, a smart contract may define the contractual conditions of a medical drug supply chain. The conditions will automatically execute upon notice of shipping of the drug from one location to another (Gupta, 2017).

A blockchain can be both permissionless (public) or permissioned (private). In a permissionless blockchain, anyone can join the network. In a private blockchain, pre-verification of the participating parties, which are all known to each other, is required. The choice between the two types is mainly driven by the use case in the particular application. If a network can "commoditize" trust, where the identity of the facilitating parties does not need to be verified, a permissionless blockchain makes sense. An example of a permissionless blockchain is Bitcoin or Ethereum. On the other hand, drug supply chain management is an ideal use case for a permissioned blockchains as it makes sense to have the participating nodes understand where in the drugs supply chain, for example, a specific drug is located. Hyperledger – an implementation of a permissioned blockchain – is an example of an open-source blockchain initiative hosted by the Linux Foundation.

6.3   HOW BLOCKCHAINS HAVE BEEN APPLIED IN THE SMART CITIES CONTEXT

Blockchain technology can contribute to smart cities in several ways. Studies have already been conducted to exploit how the association between both disruptive topics can be implemented (Ghandour et al., 2019; Xie et al., 2019; Petratos et al., 2020; Singh and Rajput, 2020). This section provides a panoramic overview of the several ways that blockchain can be an effective instrument in the smart cities context.

6.3.1   IMPROVED TRANSPARENCY AND SECURITY

This is one of the key ways in which blockchain technology can contribute to smart cities. By using blockchain, all city transactions and record-keeping can be recorded securely and transparently, allowing for greater accountability and trust in the city's operations.

One example of this is in the area of land registry and property ownership. Blockchain-based land registry systems can provide a tamper-proof record of property ownership and transactions, making it easier to verify ownership and transfer property. For example, the government of Ghana is using blockchain technology to create a tamper-proof land registry.

Another example is in the area of government procurement. By using blockchain-based systems for procurement, all bids, contracts, and payments can be recorded in a transparent and tamper-proof manner. This can help to reduce corruption and increase trust in the procurement process. For instance, the state of West Virginia in the United States implemented a blockchain-based voting system for the military overseas and it was reported that the transparency and immutability of blockchain technology helped to increase trust in the voting process.

Additionally, in the area of public transportation, blockchain technology can be used to create transparent and tamper-proof records of all transportation-related transactions, such as ticket sales and payments, which can help to increase efficiency and reduce fraud.

Blockchain technology can be used to create a transparent and tamper-proof record of energy consumption and production as well. This can help to increase the efficiency of energy distribution, reduce fraud, and enable the integration of renewable energy sources (Peise et al., 2021). For example, the Australian energy company Power Ledger is using blockchain technology to create a peer-to-peer energy trading platform that allows individuals and businesses to trade excess solar energy.

6.3.2   FACILITATE SMART CONTRACT-BASED AUTOMATION
OF CITY SERVICES AND SYSTEMS

This refers to the use of blockchain technology to automate the management and delivery of city services and systems using smart contracts. In the context of smart cities, smart contracts can be used to automate the management and delivery of city services and systems, such as transportation, energy, waste management, building management, and public services. For example, a smart contract could be used to automatically manage the payment and scheduling of public transportation services, to automatically purchase energy from a renewable energy source when it is available, or to automatically adjust the power consumption of smart appliances in response to changes in energy prices.

By using blockchain technology and smart contracts to automate the management and delivery of city services and systems, cities can increase the efficiency of their operations, reduce costs, and improve the delivery of services to citizens.

For example, the city of Singapore has implemented a blockchain-based smart parking system that uses smart contracts to automatically manage the allocation of parking spots. The system uses sensors to detect the availability of parking spots and assigns them to drivers through a smart contract. This helps to increase the efficiency of parking management and reduce traffic congestion.

The city of Amsterdam has implemented a blockchain-based smart lighting system that uses smart contracts to automatically manage the power consumption of streetlights. The system uses sensors to detect the presence of pedestrians and vehicles and adjusts the brightness of streetlights accordingly. This helps to reduce energy consumption and increase the safety of pedestrians and drivers.

The city of Shenzhen in China has implemented a blockchain-based smart city platform that uses smart contracts to automatically manage a wide range of city services, such as transportation, energy, and waste management. The platform uses a combination of sensors and cameras to collect data on city operations and uses smart contracts to automatically manage the allocation of resources and the delivery of services. This helps to increase the efficiency of city management and reduce costs.

And finally, the city of Rotterdam in the Netherlands has implemented a blockchain-based air quality monitoring system that uses smart contracts to automatically manage the monitoring and reporting of air quality. The system uses sensors to collect data on air quality and uses smart contracts to automatically report any deviations from standards to the appropriate authorities. This helps to increase the efficiency of air quality management and reduce the negative impacts of pollution on citizens' health.

6.3.3 Improved Communication and Coordination among City Departments and with Citizens

This refers to the use of blockchain technology to facilitate better communication and coordination among different city departments and with citizens.

One of the key advantages of blockchain technology is that it allows for secure and transparent communication and data sharing among different parties. In the context of smart cities, this can be used to improve communication and coordination among different city departments, such as transportation, energy, waste management, and public services. For example, a blockchain-based platform could be used to share data and coordinate actions among different city departments to improve the efficiency of city operations and the delivery of services to citizens.

Additionally, blockchain technology can also be used to facilitate better communication and coordination with citizens. For example, a blockchain-based platform could be used to provide citizens with access to real-time data on city services and systems, such as transportation schedules and energy consumption. This can help to increase transparency and trust in city operations, and can also enable citizens to provide feedback and participate in decision-making processes.

The city of Dubai has set up a pilot program using blockchain technology to improve transparency in the transportation sector. The platform allows for improved communication and coordination among different city departments involved in transportation, such as transportation planning, traffic management, and public transportation operators. It also enables citizens to access real-time data on transportation services and to provide feedback on transportation services.

6.3.4 Creation of New Economic Opportunities

This refers to the use of blockchain technology to enable the development of new decentralized applications that can create new economic opportunities for individuals, businesses, and the city.

One of the key advantages of blockchain technology is its ability to enable the creation of decentralized applications that can operate without the need for a central authority. In the context of smart cities, this can be used to create new economic opportunities by enabling the development of new decentralized applications that can provide new services and products to citizens and businesses. For example, a decentralized application could be used to create a peer-to-peer energy trading platform, or a decentralized waste management system.

The company La'Zooz; for instance, is using blockchain technology to create a decentralized ride-sharing platform that allows drivers and passengers to connect directly. The platform enables drivers to earn income by providing rides and creates new economic opportunities for the ride-sharing industry. The company OpenBazaar provides another example of using blockchain technology to create a decentralized marketplace that allows individuals and businesses to buy and sell goods and services without the need for a central authority.

SMART HEALTH AND BLOCKCHAIN – A SCENARIO

Have you ever experienced the frustration of visiting a doctor, taking exams, receiving a prescription, and then losing the paper-based prescription or discarding the exam results after some time? This situation is all too common and can lead to difficulties in the future, such as when you need to consult with another doctor and cannot remember the details of your previous medical history.

With blockchain technology, however, this problem can be solved. By utilizing a blockchain-based system for Electronic Health Records (EHRs), all of your medical history can be stored securely and accessed at any time. This includes information such as the name and specialty of any doctor you have seen, the medicines you have taken, the diagnoses you have received, and the pills you have taken and their effects.

By having all of this information in one place, you can save time, money, and paper when visiting any doctor in the future. Smart healthcare powered by blockchain technology has the potential to revolutionize the healthcare industry, and you can learn more about it in Chapter 8 of this book. Meanwhile, if you want to read more on this, check: Kassab et al. (2019). *Exploring Research in Blockchain for Healthcare and a Roadmap for the Future*. IEEE Transactions on Emerging Topics in Computing, 9(4), 1835–1852.

6.3.5 SMART EDUCATION

Educational institutions (EIs) use specialized systems to maintain course and student records, which are accessed only by staff or students on restricted websites or dedicated systems. The issuance and availability of the certificates are playing a critical role as it is strong evidence that the student has successfully completed the course (Abreu et al., 2020). However, this can lead to challenges with data security, access control, and preventing forgery. Traditional technologies can address these issues, but blockchain technology offers a more secure and trustworthy solution through verified transactions and disintermediation. Blockchain can provide a single view of students' diploma data for all interested parties, reducing the risks of information loss and preventing forged documents, making education and particularly diploma issuance even smarter. Smart education is further detailed in Chapter 10.

6.3.6 ELECTIONS

Many election systems are paper-based which can lead to a time-consuming process of counting and managing votes. Blockchain could make the election process smarter as discussed by Xie et al. (2019) and Ibrahim et al. (2021). A blockchain-based e-voting system could be structured in five steps: election creation, voter registration, vote transaction, vote tallying, and vote verification.

The election administrator creates an election using a decentralized application, authenticates and authorizes eligible voters, and assigns them a secure digital key. Voters then use this key to cast their votes, which are recorded on the blockchain and tallied automatically by the election smart contract. Each voter can verify the accuracy of the election results using the transaction ID assigned to their vote.

6.3.7 PERSONAL DATA

Blockchain can be used in a smart city to transform documentation, making documents smarter. Mobiles already help by enabling citizens to use digital versions of the physical originals. However, blockchain can leverage it by enabling the registration of unique and immutable IDs for citizens in a city blockchain, so that the citizen could easily share their data by means of the blockchain infrastructure (Xie et al., 2019). Data is currently a valuable asset in society and the economy. Then, a market where data owners can share or sell their data to data consumers can emerge. However, the potential clients have to trust an authorized third party and pay management fees to accordingly subscribe to it. To address these challenges, blockchain technology can be used to create a decentralized data exchange market where data owners and customers can cooperate without the need for an authorized third party, reducing fees and the risk of a single point of failure. Such technology could prevent recurrent risks and problems, such as document theft and forgery.

6.4 ENGINEERING SMART CITIES WITH BLOCKCHAIN

Engineering smart cities with blockchain require a series of decisions to be taken. A process can be systematically followed so that it is possible to achieve smart city

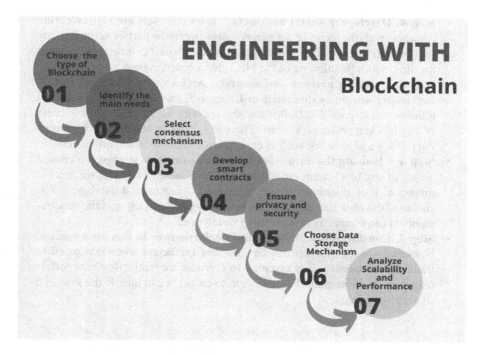

FIGURE 6.2 A process to engineer applications for smart cities using blockchain.

applications that accordingly use blockchain (Fahmideh et al., 2022). A general model of steps to be followed, as shown in Figure 6.2, includes:

- **Step 1. Choosing the type of blockchain:** The choice of the right type of blockchain depends on the specific requirements of the smart city system being developed. For instance, public blockchains may be more suitable for systems that require transparency and decentralization, while private blockchains may be better suited for systems that require greater control and privacy. For instance, the diploma issuing system for smart universities described in Section 6.3.5 probably would use a private blockchain network, since it is private to the EI.
- **Step 2. Identifying the main needs:** It's important to identify the specific needs that the blockchain-based system will support. This can help in selecting the right architecture and consensus mechanism to ensure the system is efficient and secure.
- **Step 3. Selecting the consensus mechanism:** Consensus mechanisms are critical to the security and performance of a blockchain-based system. There are different types of consensus mechanisms, as discussed in Section 6.2. The right consensus mechanism depends on the requirements of the smart city system.

- **Step 4. Developing smart contracts:** Smart contracts are self-executing contracts with the terms of the agreement between the parties being directly written into lines of code. Developing the right smart contracts is essential for the smooth functioning of the blockchain-based system.
- **Step 5. Ensuring privacy and security:** Architectural concerns, including quality attributes discussed in Chapter 3, should also be considered (Ordonez-Guerrero, 2022). Privacy and security are critical to the success of any blockchain-based system. There are different techniques for ensuring privacy and security, such as encryption and multi-factor authentication.
- **Step 6. Choosing the data storage mechanism:** Data storage is a critical aspect of any blockchain-based system. There are different storage mechanisms, such as distributed file systems and decentralized databases. The choice of the right data storage mechanism depends on the specific requirements of the smart city system being developed.
- **Step 7. Considering scalability and performance:** In certain scenarios, blockchain-based systems can be slow and inefficient when compared to traditional databases. It is essential to consider the scalability and performance of the blockchain-based system to ensure it can handle the load of a smart city system.

SHOULD I PUT EVERYTHING IN THE BLOCKCHAIN?

One common question that arises when people begin working with blockchain is: "Since the primary purpose of databases is to securely store and retrieve information, what motivates me to use blockchain instead?" The truth is that databases and blockchains are not mutually exclusive, but rather complementary technologies.

Database systems (DBS) are a traditional technology that allows engineers to store various types of data (text, personal information, images, music, videos), and use CRUD (Create, Read, Update, Delete) operations. While databases are generally trustworthy, data can be erased through the CRUD principle. In contrast, blockchain is a write-only technology that adheres to the immutability principle, making it impossible to update or delete data.

So, why is this question relevant? When dealing with data, it's crucial to consider the factors that influence data storage. Smart city systems, for example, involve a wide range of data, including security camera footage, traffic control system images, sensor data, and personal data from citizens. Storing everything in one place is not always the best strategy. For example, blockchain transactions can take longer than DBS transactions to complete, and can also be more expensive in terms of energy and/or money, depending on the consensus mechanism used. Engineers typically choose to store lightweight and sensitive data, such as personal data, in blockchains. For heavier data, a descriptive summary can be stored in blockchains, with a link to a database where the original data (video, images, etc.) can be effectively retrieved.

6.5 OPEN CHALLENGES

While blockchain technology has the potential to bring significant benefits to smart cities, several challenges need to be addressed:

- **Scalability:** One of the main challenges of using blockchain technology in smart cities is scalability. This refers to the ability of blockchain technology to handle large numbers of transactions and store large amounts of data. The current scalability of most blockchain networks is relatively low, which can be a significant challenge in the context of smart cities. As more and more devices and systems are connected to the internet and generate large amounts of data, the ability of blockchain technology to handle large numbers of transactions and store large amounts of data becomes a major concern.

    A blockchain-based platform that manages a wide range of city services, such as transportation, energy, and waste management, in a smart city may not be able to handle the high volume of data and transactions generated by a large number of connected devices and systems involved in the system at peak hours, leading to delays, interruptions, and errors in the system.

- **Interoperability:** Another challenge of using blockchain technology in smart cities is interoperability. As different city departments and organizations use different systems and platforms, it can be difficult to ensure that different blockchain-based systems can work together seamlessly. For example, a blockchain-based system that tracks and manages the usage of building systems may not be able to interoperate with other systems such as security and access control, leading to inefficiencies and difficulties in managing buildings.

- **Regulation:** As blockchain technology is a relatively new and rapidly evolving field, there is currently a lack of a clear regulatory framework for its use in smart cities. This makes it difficult for cities to implement blockchain-based systems and can also create legal and compliance risks. A city may want to implement a blockchain-based system for managing public transportation schedules and payments, but may be uncertain about how to comply with existing regulations and laws related to transportation, data privacy, and security.

- **Security and privacy:** Blockchain technology is generally considered to be highly secure, but there are still concerns about the potential for hacking and other forms of cyber attacks. This is particularly true in the context of smart cities, where large amounts of sensitive data are being stored and shared. In addition, in smart cities, blockchain technology is used to collect and share data, this can lead to privacy concerns as the personal data of citizens could be at risk of being compromised. For example, a blockchain-based system for managing public transportation schedules and payments could collect and store personal data such as passengers' travel history, location data, and payment information. This data could be vulnerable to breaches, which could lead to unauthorized access and misuse of personal data.

- **Correctness:** Some types of blockchains are based on smart contracts, a precise specification of how the transaction should be performed. Since blockchains are write-only, if there is an imprecise specification of the smart contract, this can lead to the registration of a defective block in the ledger. An option is to simulate the smart contract before effectively running it (Gomes and Coutinho, 2022). The development of simulation models for smart contracts is still something that demands further development.
- **Adoption and education:** Blockchain technology can be complex and hard to understand for some people, this can create challenges for city officials and citizens who need to be educated and trained on the technology in order to be able to use and benefit from it. So while a city may want to implement a blockchain-based system for tracking and managing the usage of building systems, the building managers and tenants may not understand how the technology works or how to use it, which could lead to low adoption rates and limited benefits.
- **Energy consumption and sustainability:** This challenge refers to the high-energy requirements of blockchain technology, which can make it difficult to implement in smart cities. Blockchain networks, especially those based on a PoW consensus mechanism, require a large amount of computational power to validate transactions and maintain the integrity of the network. This can lead to high-energy consumption and a significant carbon footprint.

6.6 FINAL REMARKS

Blockchain technology is an influential and transformative form of DLT with the potential to transform the landscape of smart cities. This chapter has demonstrated the various ways in which the technology can be utilized and modified for smart cities in various areas, such as healthcare, education, and elections. Additionally, this chapter covers a process to assist engineers in integrating blockchain solutions into smart cities. Although there are still challenges to be addressed, this chapter has explored some of them.

Starting with next chapter, we delve into specific domains of smart cities and we begin with mobility.

REFERENCES

Abreu, A. W. S., Coutinho, E. F., & Bezerra, C. I. (2020). A blockchain-based architecture for query and registration of student degree certificates. In Proceedings of the 14th Brazilian Symposium on Software Components, Architectures, and Reuse (pp. 151–160). Natal, Brazil.

Bishr, A. B. (2019). Dubai: A city powered by blockchain. Innovations: Technology, Governance, Globalization, 12(3–4), 4–8.

dos Santos Abreu, A. W., Coutinho, E. F., & Bezerra, C. I. M. (2021). Performance evaluation of data transactions in blockchain. IEEE Latin America Transactions, 20(3), 409–416.

Fahmideh, M., Grundy, J., Ahmad, A., Shen, J., Yan, J., Mougouei, D., ... & Abedin, B. (2022). Engineering blockchain-based software systems: Foundations, survey, and future directions. ACM Computing Surveys, 55(6), 1–44.

Fernandes, A., Rocha, V., da Conceição, A. F., & Horita, F. (2020). Scalable Architecture for sharing EHR using the Hyperledger Blockchain. In 2020 IEEE International Conference on Software Architecture Companion (ICSA-C) (pp. 130–138). IEEE, Salvador, Brazil.

Ghandour, A. G., Elhoseny, M., & Hassanien, A. E. (2019). Blockchains for smart cities: a survey. Security in smart cities: Models, applications, and challenges, 193–210.

Gomes, A. N., & Coutinho, E. F. (2022). Um Estudo Inicial sobre a Importância de Simular Contratos Inteligentes em Blockchain. In Anais do IV Workshop em Modelagem e Simulação de Sistemas Intensivos em Software (pp. 1–10). SBC, Salvador, Brazil.

Gupta, M. (2017). Blockchain for Dummies-IBM. IBM Limited Edition, Hoboken NJ.

Ibrahim, M., Ravindran, K., Lee, H., Farooqui, O., & Mahmoud, Q. H. (2021). Electionblock: An electronic voting system using blockchain and fingerprint authentication. In 2021 IEEE 18th International Conference on Software Architecture Companion (ICSA-C) (pp. 123–129). IEEE, Stuttgart, Germany.

Kassab, M. (2021, September). Exploring non-functional requirements for blockchain-oriented systems. In 2021 IEEE 29th International Requirements Engineering Conference Workshops (REW) (pp. 216–219). IEEE, Notre Dame, IN.

Kassab, M., DeFranco, J., Malas, T., Destefanis, G., & Neto, V. V. G. (2019). Investigating quality requirements for blockchain-based healthcare systems. In 2019 IEEE/ACM 2nd International Workshop on Emerging Trends in Software Engineering for Blockchain (WETSEB) (pp. 52–55). IEEE, Montreal, Canada.

Kassab, M., DeFranco, J., Malas, T., Laplante, P., Destefanis, G., & Neto, V. V. G. (2019). Exploring research in blockchain for healthcare and a roadmap for the future. IEEE Transactions on Emerging Topics in Computing, 9(4), 1835–1852.

Kassab, M., DeFranco, J., Malas, T., Neto, V. V. G., & Destefanis, G. (2019). Blockchain: A panacea for electronic health records?. In 2019 IEEE/ACM 1st International Workshop on Software Engineering for Healthcare (SEH) (pp. 21–24). IEEE, Montreal, Canada.

Kassab, M., & Destefanis, G. (2021, March). Blockchain and contact tracing applications for covid-19: The opportunity and the challenges. In 2021 IEEE International conference on software analysis, evolution and reengineering (SANER) (pp. 723–730). IEEE, Honolulu, HI.

Kassab, M., Destefanis, G., DeFranco, J., & Pranav, P. (2021, May). Blockchain-engineers wanted: An empirical analysis on required skills, education and experience. In 2021 IEEE/ACM 4th International Workshop on Emerging Trends in Software Engineering for Blockchain (WETSEB) (pp. 49–55). IEEE, Madrid, Spain.

Kassab, M. H., Neto, V. V. G., Destefanis, G., & Malas, T. (2021). Could blockchain help with COVID-19 crisis?. It Professional, 23(4), 44–50.

Kundu, D. (2019). Blockchain and trust in a smart city. Environment and Urbanization ASIA, 10(1), 31–43.

Ordonez-Guerrero, A. C., Munoz-Garzon, J. D., Villarreal, E. R. D., Bandi, A., & Hurtado, J. A. (2022). Blockchain architectural concerns: A systematic mapping study. In 2022 IEEE 19th International Conference on Software Architecture Companion (ICSA-C) (pp. 183–192). IEEE, Honolulu, HI.

Peise, M., Kuhlenkamp, J., Busse, A., Eberhardt, J., Ulbricht, M. R., Tai, S., ... & Zörner, T. (2021). Blockchain-based local energy grids: advanced use cases and architectural considerations. In 2021 IEEE 18th International Conference on Software Architecture Companion (ICSA-C) (pp. 130–137). IEEE, Stuttgart, Germany.

Petratos, P. N., Ljepava, N., & Salman, A. (2020). Blockchain technology, sustainability and business: A literature review and the case of Dubai and UAE. In Sustainable

Development and Social Responsibility—Volume 1: Proceedings of the 2nd American University in the Emirates International Research Conference, AUEIRC'18–Dubai, UAE 2018 (pp. 87–93). Springer International Publishing, Dubai.

Sangwan, R. S., Kassab, M., & Capitolo, C. (2020). Architectural considerations for block-chain based systems for financial transactions. *Procedia Computer Science*, 168, 265–271.

Singh, D., & Rajput, N. S. (Eds.). (2020). Blockchain Technology for Smart Cities. Springer.

Xie, J., Tang, H., Huang, T., Yu, F. R., Xie, R., Liu, J., & Liu, Y. (2019). A survey of block-chain technology applied to smart cities: Research issues and challenges. IEEE Communications Surveys & Tutorials, 21(3), 2794–2830.

7 Smart Mobility for Liveable Cities
Opportunities and Challenges

7.1 INTRODUCTION

As the world population continues to grow, cities are also expanding to accommodate more citizens, both those who were born there and those who migrate. This phenomenon has direct consequences on mobility and transportation. As the economic power of citizens increases, particularly in emerging countries, it is not uncommon to find two or more vehicles owned by a single family who lives in the same residence. This problem is intensified because the subway and train networks in several countries are still limited.

For example, according to data from the demographic census on Brazilian cities released by the Brazilian Institute of Geography and Statistics (IBGE) at the beginning of the 2010s (Brazilian Institute of Geography and Statistics, 2014), in Brazil, the subway service was present in only 20 cities of the entire territory (0.3% of the total number of Brazilian cities). The increased number of vehicles leads to increased rates of traffic congestion and frequent accidents. And such a problem is, evidently, not exclusive to developing countries.

Large metropolitan areas such as New York City or Paris also suffer from intense traffic, particularly during rush hour.

On the other hand, those who choose not to buy a car or cannot afford one, suffer from crowded transport means, scarcity of bicycle paths, and the precariousness of the services offered, a direct consequence of the increase in population with no planning or proportional increase of the urban fleets. The world also experiences a series of consequences of the global warming phenomenon, which is annually intensified due to the emission of gasses that come from the consumption of fossil fuels by the vehicle fleet. Initiatives such as carbon credit and incentives to reduce the number of vehicles daily in traffic have been imperative, not only to break global warming but also to promote a better life quality in traffic and in the urban environment as a whole.

Urban administrators have these problems on their hands and several solutions have been demanded (and implemented) around the world. This chapter presents an overview of initiatives aimed at enhancing mobility within contemporary urban areas, as well as discussing future plans for implementation in smart cities.

DOI: 10.1201/9781003348542-7

7.2 FOUNDATIONS ON URBAN MOBILITY

7.2.1 DEFINITIONS AND ENTITIES INVOLVED

Urban mobility refers to the movement of people and goods within urban areas, typically involving (and co-existing in the same space), as shown in Figure 7.1, various modes of transportation such as cars, buses, trains, bicycles, and walking. It encompasses all aspects of transportation within cities, including infrastructure, planning, and policy, as well as the integration of various modes of transportation to create a seamless and efficient system. The goal of urban mobility is to improve access and connectivity within cities, reduce traffic congestion and pollution, and promote sustainable transportation options.

Urban mobility planning is typically the responsibility of a combination of different levels of government and private entities. At the national level, government agencies such as the Department of Transportation or the Ministry of Transport are responsible for developing policies, regulations, and funding mechanisms for urban mobility. They also provide guidance and support to local governments and private entities in their efforts to improve urban mobility. At the regional level, metropolitan planning organizations (MPOs) are responsible for coordinating transportation planning and funding across multiple jurisdictions, often across multiple counties within a state. At the local level, city and county governments are typically responsible for the day-to-day management of transportation systems, including the planning,

FIGURE 7.1   Illustration of different mobility concerns in an urban environment. (1) Biking, (2) Metro, (3) Bus, and (4) Pedestrians.

design, construction, and operation of roads, public transportation, and other transportation infrastructure. They also develop and implement transportation policies, regulations, and plans.

The private sector also plays a role in urban mobility planning. Private companies, such as transportation service providers, technology companies, and consulting firms, often work with government agencies to provide services and expertise related to urban mobility planning, such as planning, engineering, and technology development. Overall, urban mobility planning is a collaborative effort that involves different levels of government and private entities working together to create a transportation system that is safe, efficient, and sustainable.

7.2.2 ROAD PLANNING

Road planning plays a crucial role in urban mobility because it determines the layout and infrastructure of a city's transportation system. A well-planned road network can improve access and connectivity within a city, reduce traffic congestion and travel times, and promote sustainable transportation options such as walking, cycling, and public transportation. Proper road planning involves considering a wide range of factors such as population density, land use, economic activity, and environmental concerns. It also involves designing roads to accommodate different modes of transportation, such as cars, buses, bikes, and pedestrians, and creating a balance between them. This can include building dedicated bike lanes and sidewalks, creating intersections that are safe and easy to navigate for all users, and designing roads to encourage slower speeds in areas with high pedestrian activity.

Additionally, road planning also includes creating a road network that is resilient to natural disasters and climate change as well as reducing dependency on fossil fuels. In recent years, cities are turning toward sustainable transportation options and smart city solutions to improve urban mobility.

7.2.3 TECHNOLOGIES

Several technologies are involved in urban mobility planning, including:

* **Geographic Information Systems (GIS):** GIS technology is used to map and analyze transportation data, such as traffic flow, population density, and land use patterns. This information is used to support transportation planning and decision-making.
* **Traffic simulation:** Traffic simulation software is used to model and predict traffic flow, traffic congestion, and the impact of different transportation projects on the city's transportation system. Several traffic simulators exist, such as OpenTrafficSim[1] and Traffic3D, as discussed in Section 4.3.
* **Intelligent Transportation Systems (ITS):** ITS technology is used to improve the efficiency and safety of the transportation system. This can include traffic monitoring systems, traveler information systems, and advanced transportation management systems.

- **Connected and Automated Vehicles (CAVs):** CAVs technology allows vehicles to communicate with each other and with infrastructure, such as traffic signals, to improve traffic flow and safety.
- **Big Data Analytics:** Big data analytics are used to process and analyze large amounts of data from various sources, such as GPS, social media, and sensor networks, to gain insights into transportation patterns and behavior.

These technologies are used to support transportation planning and decision-making, improve the efficiency and safety of the transportation system, and promote sustainable transportation options.

7.2.4 PUBLIC TRANSPORTATION

Public transportation is a key component of urban mobility planning because it plays a crucial role in providing access and connectivity within a city. Urban mobility planning aims to create a transportation system that is safe, efficient, and sustainable, and public transportation is an important part of achieving this goal.

Here are a few ways in which public transportation is related to and impacted by urban mobility planning:

- **Network planning:** Urban mobility planning involves designing a transportation system that is efficient and provides good connectivity within the city. This includes designing a public transportation network that is easy to use and covers all areas of the city, with frequent and reliable service.
- **Integration with other modes of transportation:** Urban mobility planning aims to create a seamless transportation system that integrates different modes of transportation, such as cars, bikes, and walking. This includes designing public transportation systems that integrate with other transportation options, such as bike-sharing and ride-hailing services.
- **Investment in infrastructure and technology:** Urban mobility planning involves investing in the infrastructure and technology needed to support public transportation, such as buses, trains, and subway systems. This includes upgrading existing systems and building new infrastructure to meet the growing demands of the city's population.
- **Encouraging the use of public transportation:** Urban mobility planning involves encouraging the use of public transportation as an alternative to private cars. This can include implementing policies such as congestion charging, dedicated bus lanes, and providing real-time information to riders.

7.2.5 SUSTAINABILITY AND URBAN MOBILITY PLANNING

Sustainability is closely related to urban mobility because the way we move around cities has a significant impact on the environment and our quality of life. Urban transportation is a major source of greenhouse gas emissions, air pollution, and noise pollution. It also contributes to traffic congestion, which can reduce economic productivity and negatively impact the health and well-being of city residents.

To promote sustainability in urban mobility, cities are implementing a variety of strategies such as:

- Encouraging the use of public transportation, walking, and cycling as alternatives to private cars.
- Investing in electric and low-emission vehicles and public transportation.
- Developing infrastructure and policies to support active transportation, such as building dedicated bike lanes and sidewalks.
- Implementing traffic management strategies to reduce congestion and improve traffic flow.
- Promoting carpooling and ride-sharing.
- Investing in smart transportation systems that use technology to improve the efficiency and sustainability of the transportation system.

Promoting sustainability in urban mobility is essential for creating livable, healthy, and prosperous cities, and it can be achieved by encouraging the use of sustainable transportation options, investing in sustainable transportation infrastructure, and implementing smart city solutions.

7.3 PUBLIC GREEN INITIATIVES AND UNDERPINNING TECHNOLOGIES

In this section, we discuss the strategies frequently used in urban areas to promote life quality while leveraging sustainability.

In the effort to create more sustainable and environmentally friendly cities, many governments have implemented green initiatives. Herein, *green* is the term used to denote practices that reinforce sustainability and environmental benefits, such as decreasing carbon emissions or intrusive interventions. These initiatives, whether utilizing technology or not, aim to reduce the number of vehicles on the roads, decrease emissions and traffic congestion, and improve the quality of air in cities.

Green initiatives in the urban mobility context are frequently related to encouraging the reduction of individual vehicle use, public policies to reward carpooling/ride-sharing, and the development of infrastructure to enable/encourage the use of bicycles and walks, as shown in Figure 7.2, creating a systemic benefit to the environment and the health of users.

Figure 7.2 shows a ranking of how sustainable the common practices currently found in Urban Mobility Planning are. Walking is certainly in the first place. Perhaps the most widely adopted initiative is promoting the development of communities that are designed to be easily walkable and bikeable, with safe sidewalks and bike[2] lanes, to encourage active transportation. An example of this initiative is the development of communities like Vauban in Freiburg, Germany, which is a car-free neighborhood designed to be easily walkable and bikeable, with safe sidewalks and bike lanes.

FIGURE 7.2   A ranking of sustainable practices in urban mobility.

Another example is the development of the Ciclovia program in Bogota, which closes certain streets to cars on Sundays and holidays, creating a safe space for people to walk and bike. Bicycles are not only an environmentally friendly option but also a necessary means of transportation in high population density areas like China and India. Electric bicycles have been identified as a low-carbon transport system of the future and public bike sharing has seen a resurgence in popularity in some countries. During the COVID-19 pandemic, bike sharing was also seen as a healthy transportation option, and its usage increased among unemployed people due to the low cost. Bicycles have received significant encouragement for their use due to the appeal of studies that bring impactful numbers.

In the second place, we find carpooling and ride-sharing programs; encouraging the sharing of vehicles among multiple people to reduce the number of cars on the road and decrease emissions (such as BlaBlaCar[3]). This topic is further explored in Section 7.4. An example of this initiative is the implementation of carpool lanes on highways, where only vehicles with multiple occupants are allowed to use them. Another example is the development of ride-sharing apps such as Uber and Lyft, which allow individuals to share rides and reduce the number of cars on the road.

Plate rotation policy is also a co-related practice, as the one in São Paulo, Brazil. Under this policy, certain vehicles are prohibited from driving in specific regions for six hours one day a week. This policy aims to decrease the number of cars on the road and reduce emissions.

ARE METROS SUSTAINABLE MEANS?

Metro trains (subway) can be a sustainable means of transportation because they can help reduce greenhouse gas emissions, improve air quality, and reduce traffic congestion.

However, deploying a metro system can come with significant costs and environmental impacts, particularly during the construction phase. For example, building underground tunnels and stations can be expensive and may require significant excavation and disruption of the surrounding environment. Additionally, the manufacture of materials such as concrete and steel for the construction of metro systems can generate significant greenhouse gas emissions.

There may also be trade-offs between the sustainability benefits of a metro system and other social and economic considerations. For example, a metro system may provide efficient and sustainable transportation options, but it may also displace communities or disrupt local businesses during construction. Additionally, the cost of building and operating a metro system may be high, which could lead to affordability issues for low-income populations.

The third place is the initiative of Electric and hybrid vehicle incentives; offering tax breaks, rebates, and other financial incentives to encourage the purchase of electric and hybrid vehicles. An example of this initiative is the federal tax credit for electric vehicles in the United States, which provides a tax credit of up to $7500 for the purchase of a qualifying electric vehicle. Another example is the state of California's Clean Vehicle Rebate Project, which offers rebates of up to $7000 for the purchase or lease of a qualifying electric or hybrid vehicle.

Collective transport means come in fourth place since they occupy a relevant space, exhibit carbon emission, but transport many people at once. In this category, Metro Trains could come first then buses, since they are majorly electric. Finally, the worst for the urban environment is individual vehicles, which are not even listed in the ranking.

WHAT ABOUT ELECTRIC CARS?

Electric cars are a reality and they are becoming increasingly popular. Electric cars are powered by electric motors that run on rechargeable batteries, rather than by internal combustion engines that run on gasoline or diesel.

The sustainability of electric cars depends on a number of factors, including the sources of electricity used to charge them, the materials used to manufacture the batteries, and the environmental impact of battery disposal. In general, electric cars have the potential to be more sustainable than gasoline or diesel-powered cars, as they produce lower greenhouse gas emissions and air pollutants during use. One key factor in the sustainability of electric cars is the source of the electricity used to charge them. If the electricity comes from renewable sources such as wind or solar power, electric cars can be a very sustainable option. However, if the electricity comes from fossil fuels such as coal or natural gas, the sustainability benefits of electric cars may be reduced.

The production and disposal of batteries can also have environmental impacts. The production of lithium-ion batteries, which are commonly used in electric cars, requires the extraction and processing of raw materials such as lithium, cobalt, and nickel, which can be environmentally intensive processes. Additionally, the disposal of batteries at the end of their life cycle can pose environmental challenges if not handled properly.

There are many examples of commercial electric cars available in the market, including the Tesla Model 3, Nissan Leaf, Chevrolet Bolt EV, BMW i3, Hyundai Kona Electric, and Ford Mustang Mach-E, available in many countries worldwide.

Table 7.1 brings numbers to illustrate that. Consider the space occupied by people by walking, using a bicycle, using public transport, and using cars. It is known that one square meter (m^2) can be occupied by nine people. Then, 5 m^2 could accommodate 45 citizens. In comparison, a bus, which regularly occupies around 50 m^2, transports an average of 48 passengers per trip. One single bike occupies around 1.91 m^2. Forty-five bikes with citizens would occupy around 86 m^2. A single car occupies between 6 m^2 and 11 m^2 (not considering the need for space around it). If each car has only one person inside, 45 citizens would take between 270 m^2 and 495 m^2 in a conservative estimation. These data show how beneficial can be to share a car. If five people share a single car, the space occupied could be reduced to something close to an entire bus. Even without performing a quick trade-off analysis on these data about sustainability and urban space occupied, it's known that using bikes as a mode of transportation is more sustainable because there are no carbon emissions. In terms of space usage, bikes are also well-ranked compared to buses and cars. Bikes only occupy 1.91 m^2 per person (86 m^2 for 45 citizens), while buses occupy 50 m^2 per 48 passengers (but emit gases, harnessing sustainability) and cars occupy between 6 m^2 and 11 m^2 if driven by a single person. Then, bikes are a good option, since a walking path can be exhausting or too long, buses can be a good collective option

TABLE 7.1
Comparison between the Average Space Occupied by 45 Citizens

	By walk	In Bikes	In a bus	Using individual cars	5 people sharing a single car
Space occupied by 45 citizens (in m^2)	5 m^2	86 m^2	50 m^2	270–495 m^2	54–99 m^2

(but emits carbon) and cars are ultimately the worst option. The impact of cars can be reduced by sharing them. However, it is worth noting that while bikes are more space-efficient and sustainable, they may not be as practical or suitable for certain areas or types of transportation, which demands the creation of policies and bike lanes to make them feasible.

To encourage the widespread adoption of bicycles, governments have invested in creating or improving bicycle lanes and paths. For example, Brazil's National Urban Mobility Policy prioritizes non-motorized means of transportation, such as bicycles, and emphasizes the integration of urban development policies with transportation policies (Benedini et al., 2020). In Vannes, France, citizens can borrow a bicycle from the city government for a year.

Green initiatives and policies aimed at promoting bicycle use. Decreasing the number of vehicles on the road has the potential to improve the environment and livability of cities. However, for these initiatives to be effective, they must be accompanied by appropriate infrastructure and technological support to aid citizens in making informed transportation choices. In fact, technologies such as mobile apps have been developed for this purpose. One category of these apps provides information on the feasibility and benefits of using bicycles instead of cars or public transportation. For example, a project developed as a cooperation between the IntersCity group[4] based on the Department of Computer Science of the University of São Paulo in Brazil and the Senseable City Lab from the Massachusetts Institute of Technology in the United States, named BikeScience (http://bikescienceweb.interscity.org/), offers an analytical method that processes millions of bike-sharing trips and analyze bike sharing mobility, abstracting feasible mobility flows across specific urban areas that can support urban policymakers with planning decisions (Kon et al., 2022). In that context, a feasible mobility flow is a term used to describe the urban paths that can be taken by bikers to move from one place to a certain destination. A feasible mobility flow takes into account factors such as the physical layout of the city, population density, and the availability and accessibility of different modes of transportation. The accessibility of the road is also considered, for instance, to recommend a path for a biker.

In Noronha et al. (2022), a tool was developed to evaluate urban routes and identify streets that could benefit from retrofitting. Retrofitting in urban routes refers to the process of modifying existing infrastructure to improve its functionality, safety, or sustainability. This can include upgrading existing roads, sidewalks, and bike lanes to make them more accessible and safe for pedestrians and bicyclists, or installing new public transportation systems such as buses or trains. Retrofitting can also include adding new features to existing infrastructure, such as adding bike-sharing stations or electric vehicle charging stations. The goal of retrofitting is to make urban routes more efficient, sustainable, and accessible to all users, whether they are walking, biking, or driving. It can also include upgrading buildings, homes, and neighborhoods to make them more energy-efficient and resilient to the effects of climate change, which is called retrofitting for energy efficiency and climate resilience. Noronha and colleagues' approach aims to make walking easier, and safer, and reduce car dependence by analyzing streets and suggesting interventions that can improve mobility in urban areas. Their tool identifies candidate streets for retrofitting, creates prototypes, and quantitatively assesses their potential to improve mobility, providing policymakers with valuable information to design new transportation solutions.

7.4 REWARDING MECHANISMS

Rewarding policies are another way in which smart cities can promote sustainable mobility. These policies incentive citizens to take actions that benefit the city's transportation system, such as giving rides to others or using their own vehicles to provide transportation during times when public buses are not running.

One example of a reward policy is Waze Carpool, which rewards drivers with money for giving rides to others. The program, which was launched in 2018 and ended in September 2022, aimed to optimize the use of cars that are already on the road, rather than adding more vehicles to the streets. Another possible solution is to reward drivers with discounts on taxes to reduce recurrent problems in traffic. Graciano-Neto et al. (2020) discussed a scenario that could be implemented, as shown in Figure 7.3. The figure shows a virtuous circle that can be established in a mobility system of a smart city with private-public partnerships and government policies. In their study, the authors discuss how several dimensions of a smart city ecosystem should interplay to turn technological solutions such as the one discussed in reality. The figure depicts a situation in which cars (autonomous, as discussed in the next section, or not) could join the smart city infrastructure (that includes a platform and network) and offer novel services for the population, such as a ride for people when the buses are too crowded. This could be rewarded by a discount given by the fuel station because the municipal

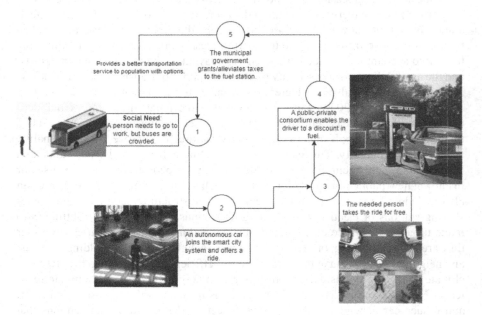

FIGURE 7.3 A scenario in which a reward mechanism could be implemented for mobility. (Adapted from Graciano-Neto et al. 2020.)

government also gave a discount on taxes to the fuel station to reimburse its discount to the driver that improved the quality of city mobility for the population. This is a simplistic scenario, but that can be even more complex, and tangled and involve many other actors, firms, private and public institutions, and members of society that could offer services for the city. Providing exchanges of services and reimbursements as illustrated is possible via information systems that deal with and process the involved information to provide the intended behavior. However, the illustrated situation is only possible if social, business and technical dimensions of the smart city ecosystem are synergic, that is, since a social need (the need of citizens to arrive at their work damaged by the overcrowded public buses) can be solved in the technical perspective (a new system that joins the city can support that need and solve the problem) by the establishment of crossed advantages via business alliances between companies (the public sector and private companies) and the citizens as well.

Additionally, the use of autonomous cars could also be rewarded, as they are one of the main solutions discussed to improve mobility in cities. Section 7.5 further explores the topic of autonomous cars and their potential impact on smart cities.

7.5 TECHNOLOGICAL SOLUTIONS

Several technological solutions can be used to improve mobility in smart cities:

1. **Autonomous vehicles:** Self-driving cars, buses, and trucks can reduce the need for human drivers, improve traffic flow, and increase the efficiency of transportation systems. Waymo, a subsidiary of Alphabet (Google's parent company), has been testing autonomous vehicles on public roads in Phoenix, Arizona since 2017. They have also started a commercial ride-hailing service using autonomous vehicles. Another example is Tesla, the company has been releasing updates to its cars to make them more autonomous and have a feature called "full self-driving" which is currently in beta testing. We delve into the topic of autonomous vehicles in the next section.

2. **Smart traffic control systems:** These systems use real-time data and machine learning algorithms to optimize traffic flow, reduce congestion, and improve safety. The city of Singapore has implemented a system called ITS to optimize traffic flow and reduce congestion. The system uses cameras and sensors to gather real-time traffic data, which is then analyzed by an algorithm to adjust traffic lights and manage traffic flow.

3. **Intelligent transportation systems (ITS):** These systems use information and communication technologies to improve transportation efficiency, safety, and sustainability. Examples include electronic toll collection systems, real-time traffic information systems, and advanced traveler information systems. Many cities around the world have implemented electronic toll collection systems, such as E-ZPass in the northeastern United States and SunPass in Florida. These systems allow drivers to pay tolls electronically, reducing the need for cash payments and speeding up the toll collection process.

4. **Connected vehicles:** These vehicles use wireless communication technologies to share information with other vehicles and with transportation

infrastructure, such as traffic lights and road signs. This can improve traffic flow, reduce accidents, and provide drivers with real-time information about traffic conditions. The city of Wuhu, China has implemented a connected vehicle system that uses the Dedicated Short-Range Communications (DSRC) standard to connect vehicles with traffic lights, signs, and other vehicles. This allows vehicles to share information about traffic conditions and improve traffic flow.

5. **Public transportation optimization:** The use of real-time data, route optimization algorithms, and mobile ticketing can improve the efficiency and convenience of public transportation systems. The city of Curitiba, Brazil has implemented a bus rapid transit system that uses dedicated bus lanes and optimized routes to improve the efficiency and speed of public transportation.

6. **Bike-sharing and car-sharing programs:** These programs allow individuals to rent bicycles or cars on demand, reducing the need for personal vehicle ownership and encouraging the use of more sustainable transportation options. Bike-sharing programs, such as Citi Bike in New York City and Capital Bikeshare in Washington, D.C., have been implemented in many cities around the world. Car-sharing programs, such as Zipcar and Car2Go, have also been implemented in many cities, allowing individuals to rent cars on demand.

7. **Smart parking:** Smart parking systems use sensors, cameras, and mobile apps to help drivers find available parking spaces and reduce the time and fuel consumption associated with searching for a parking spot. The city of San Francisco has implemented a smart parking system that uses sensors to detect available parking spaces and provides real-time information to drivers through a mobile app.

8. **Multimodal trip planning:** Smart trip planning apps and platforms allow individuals to plan and book trips using multiple modes of transportation, such as buses, trains, bikes, and cars. This can make it easier for people to find the most efficient and sustainable transportation options for their needs. Many cities around the world have implemented multimodal trip-planning apps, such as Google Maps and Citymapper, that allow individuals to plan trips using multiple modes of transportation, including public transportation, biking, and walking.

9. **Electric vehicle charging infrastructure:** The deployment of electric vehicle charging stations can support the adoption of electric vehicles, which can help to reduce pollution and greenhouse gas emissions from the transportation sector. Any countries and cities around the world have begun to implement electric vehicle charging infrastructure, such as the installation of charging stations along highways and in public parking lots. For example, in the United States, the company ChargePoint has installed over 113,000 charging spots across North America, Europe.

10. **MaaS (Mobility as a Service):** MaaS platforms integrate different transportation options and services, such as public transportation, bike-sharing, and ride-hailing, into a single platform, making it easy for people to plan, book, and pay for their trips.

7.6 AUTONOMOUS VEHICLES

Autonomous vehicles are widely regarded as one of the most highly anticipated technological advancements in recent times. An autonomous car as a constituent of an Urban Mobility System (another expression for Smart Traffic Control System) can contribute with several capabilities and possible scenarios, such as (Pelliccione et al. 2016): Parking place search, reporting road conditions, vehicle platooning, cooperative collision avoidance, providing basic information (location, heading, speed, and acceleration) that can then be used for many different types of services. In addition, a car can contribute directly to urban mobility, through collaborative rides practices encouraged by the city government (Teixeira et al. 2020).

Currently, there are several collaborative practices, part of the phenomenon of a global shared economy, that have a key connection to collaboration on sharing resources or services that are being underused. One such practice is the project Carro Leve (Light Car, in a free translation) in Brazil (Castañé, 2015). This project is an electric car-sharing system, and the system has been operational with three electric cars. In addition to contributing to the improvement of urban mobility through car sharing, the project also encourages cultural changes, as the price fare encourages the sharing of rides during each trip.

The first example of how autonomous cars could help in traffic is brought by Teixeira et al. (2020). In their study, authors investigated the degree of operational independence (see Chapter 3 on systems of systems), that is, how a system (in this case, an autonomous car) that is part of a smart city decides whether it should keep the accomplishment of its own goals (such as conducting the owner to a final destination) or choose to have a small drift in its path to give a ride to someone and receive some reward, using Waze or similar apps to receive money, carbon credit or public taxes discounts.

In their study, researchers examined how an autonomous vehicle would handle requests for rides based on its level of operational independence. Through simulation, they evaluated different levels of acceptance for ride requests, ranging from 10% to 60%. They also considered scenarios in which the smart city system requests emergency rides, such as transportation to a hospital. In these cases, it was determined that the car's user, rather than the car's system, should be responsible for making decisions on whether to accept these requests. The simulation results showed that the number of rides accepted by the car closely matched the established level of operational independence. Specifically, when the car system received 356 ride requests, it accepted 42 of them, which equates to an acceptance rate of 11.79%. Additionally, the results for emergency ride requests were consistent with expectations. Overall, the research highlights the importance of allowing systems in a smart city to make autonomous decisions to maintain operational independence.

Autonomous cars, also known as self-driving cars, have been developed using several technologies. According to Oquendo (2019a, 2009b), a platoon is a group of cars that travel together at a stable fixed velocity to maintain a safe distance and improve traffic flow. Platoons can be formed in two ways: By individual self-driving cars that control their velocity and distance to surrounding vehicles, or by connected self-driving cars that communicate with each other to coordinate movements, forming what is known as an Internet-of-Vehicles. This concept is particularly useful in Smart Traffic Control Systems.

NOT THAT AUTONOMOUS!

Uber had plans to begin using driverless vehicles for food delivery in 2021. However, concerns have been raised about the safety of autonomous cars, as evidenced by accidents reported in the literature and news. For example, a study published in 2020 found that 300 traffic accidents involving autonomous vehicles occurred in 46 locations in California (Petrović et al., 2020). These incidents suggest that the technology still needs to be improved before it can be used for commercial and industrial applications. Tesla's cars, which rely solely on computer vision sensors, have been involved in several accidents. In some cases, the sensors were unable to distinguish hazard lights from other vehicles stopped in the lane, resulting in accidents involving first responders (Zaparolli, 2022).

Despite the challenges, there have also been significant and successful advancements in autonomous vehicle technology in recent years. For example, Scania invested $230,000 in a project to develop a robotic truck in partnership with researchers from the University of São Paulo in São Carlos, Brazil. The project, which began in 2013, used the G360 model, a 9-ton truck, and required mechanical changes, the installation of sensors, and electronic systems, including a stereo camera system and a high-precision GPS device. Three cameras installed in front of the truck were used to estimate the distance from objects in the image, similar to how human eyes work (da Silveira, 2022. Another research project, called Intelligent Robotic Car for Autonomous Navigation (Carina), began in 2011 to demonstrate that low-cost technology could be developed. In October 2013, the vehicle was tested on the streets of São Carlos without a driver.

Artificial Intelligence has also been employed to empower autonomous cars (Silva et al. 2022) and simulations have been adopted to evaluate self-driving car technologies (Santana et al., 2021). These technologies can help autonomous cars to be an effective part of Smart Traffic Control Systems.

Figure 7.4 illustrates part of a typical Smart Traffic Control System. In those systems, we expect that most of the urban mobility elements can have sensors and software: the crosswalks, the autonomous cars, the traffic lighting systems, and all other elements. We expect that the emergent behaviors achieved can be a result of the communication established between all these technologies. For instance, fluid traffic can be achieved as the result of the platoon among the connected self-driving cars, but also as a result of the communication with the traffic light system that synchronizes its signals with others and with the cars. Simulation technology can be used to predict an ideal speed that makes the traffic flow and the entire system can adjust its velocity so that the emergent behavior of fluid traffic is achieved. The behavior safe traffic can also be achieved not only as a result of the platooning but also from the interoperability between the crosswalk sensors, which can detect a pedestrian presence and communicate with the traffic lightning system to turn red and also

FIGURE 7.4   An illustration of traffic with sensors and presence detection mechanisms.

notify autonomous cars that they should stop, avoiding collisions because of eventual failures in cars sensors.

Electric cars have become common and this is considered a cleaner energy consumer. The air can also be exploited. Prototypes of flying vehicles have also been seen around the world. But the aerial movement solutions are still prototypes and raise other discussions: When this happens, should the person have a driver's license or a pilot's license? or both? Several technologies have been developed to make urban mobility not only more efficient but also more sustainable. The next section discusses the social impact of these technologies.

7.7   SOCIAL IMPACT

Several of these technologies can bring important benefits for mobility and the population at large. However, some social issues have to be discussed so that technology can be not only innovative but also democratic.

From a legal point of view, liability is an important issue, that is, who would respond if an accident occurs due to a driverless car, for instance. The car's owner? The fabricant?

Another important issue concerns the rewarding mechanisms: Who would benefit from rewarding policies? Those who have the economic power to have a car and/or a bike? Inclusive public policies are also demanded to be developed in alignment with the technologies so that such technologies can reach the entire population. Examples of it could be the development of bike lanes, sensor distribution, and bicycles donation or loans.

Using private cars to solve problems when an accident occurs or a bus breaks down can be interesting, but we could also think that this is a statement that the public service offered is inefficient and a symptom that maintenance should be more frequent. From such a perspective, technologies would be palliatives for systemic problems that should be studied and for which more efficient solutions should be conceived. All these aspects should also be thought and planned so that the forthcoming mobility technologies can be inclusive to all social classes and be effective, not be distractions to the inefficiency of public management.

Smart mobility has the potential to bring several social benefits to smart cities. Some of these benefits include:

- **Increased accessibility:** Solutions, such as ride-sharing and bike-sharing programs, can improve accessibility for people who are unable to drive or do not have access to a vehicle.
- **Improved traffic flow:** Smart traffic management systems can help to reduce congestion and improve traffic flow, resulting in shorter travel times and less pollution.
- **Increased safety:** Autonomous vehicles and other smart mobility solutions can help to reduce the number of accidents caused by human error, increasing safety on the roads.
- **Enhanced quality of life:** Mobility systems can help to improve the quality of life for citizens by reducing noise and air pollution, providing more efficient and convenient transportation options, and increasing access to job opportunities.
- **Social equity:** Smart mobility systems can help to increase social equity by providing more accessible transportation options for low-income and marginalized communities.
- **Reduced costs:** The conceived solutions can help to reduce transportation costs for individuals and households, particularly for those who rely on personal vehicles for transportation.
- **Increased environmental sustainability:** Smart solutions can help to reduce greenhouse gas emissions, air pollution, and dependence on fossil fuels, contributing to overall environmental sustainability.

However, to make it work, the role of government in the establishment of public policies to foster smart mobility in smart cities is crucial. Some of the key ways in which government can support the development of smart mobility include:

- **Investment in infrastructure:** Public administration can provide funding for the development of infrastructure such as sidewalks, bike lanes, and public transportation systems that support smart mobility.
- **Regulation and standards:** Government could establish regulations and standards for the deployment of smart mobility technologies to ensure safety and interoperability.
- **Data sharing and management:** Public managers can support the collection and sharing of data from transportation systems to support smart mobility decision-making.

- **Encouraging innovation:** Governmental initiatives can provide incentives for companies and researchers to develop and deploy innovative smart mobility solutions.
- **Public–private partnership:** Government can partner with private companies and organizations to develop and implement smart mobility solutions.
- **Education and awareness:** Public policies can support education and awareness programs to educate citizens about the benefits and opportunities of smart mobility.
- **Inclusion and Equity:** Government can also ensure that smart mobility policies and solutions are inclusive and equitable, addressing the needs of marginalized and low-income communities.
- **Research and development:** Public initiatives could support research and development of new technologies and business models to foster smart mobility and reduce barriers to adoption.

Overall, the government can play an important role in fostering smart mobility by providing funding, creating regulations and standards, and supporting research and development. By doing so, it can help to create an enabling environment for the development and deployment of smart mobility solutions, contributing to the overall goal of making smart cities more livable, sustainable, and equitable.

7.8   FINAL REMARKS

In this chapter, we have seen that many of the technological solutions developed to improve urban mobility focus on reducing car usage, through public policies that incentive shared rides and the development of infrastructure to support walking and biking, leading to systemic benefits for both the environment and public health. Additionally, routing technologies for cars are commonly implemented, but smart traffic systems and autonomous vehicles are still in the prototyping stage. Therefore, it is important to not only address immediate sustainability concerns by implementing policies and infrastructure that promote green transportation but also to prepare for the future by building technology-intensive infrastructure that will allow for even greater optimization in the years to come, as smart traffic becomes a reality. The next chapter discusses how health is covered in smart cities.

NOTES

1 https://opentrafficsim.org/
2 Herein, the terms "bike" and "bicycle" will be interchangeably used.
3 https://www.blablacar.com/
4 https://interscity.org/

REFERENCES

Benedini, D. J.; Lavieri, P. S., Strambi, O. Understanding the use of private and shared bicycles in large emerging cities: The case of Sao Paulo, Brazil. Case Studies on Transport Policy, v. 8, n. 2, p. 564–575, 2020.

Boscarioli, C., Araujo, R.M. & Maciel, R.S.P. (2017). I GranDSI-BR – Grand Research Challenges in Information Systems in Brazil 2016 – 2026. Special Committee on Information Systems (CE-SI). Brazilian Computer Society (SBC). 184p ISBN: [978-85-7669-384-0]

Brazilian Institute of Geography and Statistics (IBGE). "Perfil dos municípios brasileiros: 2013" (Profile of Brazilian municipalities: 2013). ISBN: 9788524043208. 282 pages. Available at: https://biblioteca.ibge.gov.br/visualizacao/livros/liv86302.pdf (In Portuguese).

Castañé, A. (2015) "Cidades inteligentes, práticas colaborativas." Revista Fonte Tecnologia da Informação na Gestão Pública 12(15). 39–40. Prodemge.

Cavalcante, E., Cacho, N., Lopes, F. & Batista, T. (2017). Challenges to the development of smart city systems: A system-of-systems view. In Proceedings of the Brazilian Symposium on Software Engineering (pp. 244–249). Fortaleza CE, Brazil

da Silveira, E. (2022). O futuro sem motorista (The driverless future). Revista FAPESP. Edition 235. Available at: <https://revistapesquisa.fapesp.br/o-futuro-sem-motorista/>

Delécolle, A., Lima, R. S., Graciano-Neto, V. V., & Buisson, J. (2020). Architectural strategy to enhance the availability quality attribute in system-of-systems architectures: A case study. In Proceedings of the System of Systems Engineering Conference 2020 (pp. 93–98). Budapest, Hungary.

Graciano-Neto, V. V., Manzano, W., Kassab, M. & Nakagawa, E. Y. (2018). Model-based engineering & simulation of software-intensive systems-of-systems: Experience report and lessons learned. In Proceedings of the European Conference on Software Architecture (Companion) (27:1–27:7). Madrid Spain.

Graciano-Neto, V. V., Santos R.P., Viana D. & Araujo R. (2020). Towards a conceptual model to understand software ecosystems emerging from systems-of-information systems. In: Santos R., Maciel C., Viterbo J. (eds) Software Ecosystems, Sustainability and Human Values in the Social Web. WAIHCWS 2017, WAIHCWS 2018. Communications in Computer and Information Science, vol 1081. Springer, Cham. https://doi.org/10.1007/978-3-030-46130-0_1.

Gu, T., Kim, I., & Currie, G. (2021). The two-wheeled renaissance in China—An empirical review of bicycle, E-bike, and motorbike development. International Journal of Sustainable Transportation, 15(4), 239–258.

Guo, X., Lu, C., Sun, D., Gao, Y., & Xue, B. (2021). Comparison of usage and influencing factors between governmental public bicycles and dockless bicycles in Linfen City, China. Sustainability, 13(12), 6890.

Jobe, J., & Griffin, G. P. (2021). Bike share responses to COVID-19. Transportation Research Interdisciplinary Perspectives, 10, 100353.

Kon, F., Ferreira, É.C., de Souza, H.A. et al. (2022) Abstracting mobility flows from bike-sharing systems. Public Transport 14, 545–581. https://doi.org/10.1007/s12469-020-00259-5.

Martins, T. G., Lago, N., Santana, E. F., Telea, A., Kon, F., & de Souza, H. A. (2021). Using bundling to visualize multivariate urban mobility structure patterns in the São Paulo metropolitan area. Journal of Internet Services and Applications, 12(1), 1–32.

Noronha, M., Celani, G., & Duarte, J. P. (2022). Computationally evaluating street retrofitting interventions. Nexus Network Journal, 24 (2), 481–502.

Oquendo, F. (2019a). Architecting systems-of-systems of self-driving cars for platooning on the internet-of-vehicles with SosADL. IFIPIo, 3–20

Oquendo, F. (2019b). Architecting exogenous software-intensive systems-of-systems on the internet-of-vehicles with SosADL. Systems Engineering, 22(6), 502–518.

Pelliccione, P., Kobetski, A., Larsson, T., Aramrattana, M. T., Aderum, S. M. Agren, Jonsson, G., Heldal, R., Bergenhem, C., & Thorsén, A. (2016). "Architecting cars as constituents of a system of systems," In Proceedings of SiSoS@ECSA (p. 5). ACM, Copenhagen, Denmark.

Pereira, G. A. S., Drews-Jr, P. L. J., Wolf, D. F., & Mattos, L. S. (2021) ICAR 2019 special issue. Journal of Intelligent and Robotic Systems, 102(4), 88.

Petrović, D., Mijailović, R., & Pešić, D. (2020). Traffic accidents with autonomous vehicles: Type of collisions, manoeuvres and errors of conventional vehicles' drivers, Transportation Research Procedia, 45, 161–168, https://doi.org/10.1016/j.trpro.2020.03.003.

Santana, E. F. Z., Chaves, A. P., Gerosa, M. A., Kon, F. & Milojicic, D. S. (2018). Software platforms for smart cities: Concepts, requirements, challenges, and a unified reference architecture. ACM Computing Surveys, 50(6), Article 78, 37 pages. https://doi.org/10.1145/3124391

Santana, E. F. Z., Covas, G., Duarte, F., Santi, P., Ratti, C., & Kon, F. (2021). Transitioning to a driverless city: Evaluating a hybrid system for autonomous and non-autonomous vehicles. Simulation Modelling Practice and Theory. 107, 102210.

Silva, C. E., César, T. S., Gomes, I. P., Silva, J. A. R., Wolf, Alvers, R., Souza, J.. (2022). Route scheduling system for multiple self-driving cars using K-means and bio-inspired algorithms. In Iliadis, L., Jayne, C., Tefas, A., Pimenidis, E. (eds) Engineering Applications of Neural Networks.EANN 2022. Communications in Computer and Information Science, vol 1600. Springer, Cham. https://doi.org/10.1007/978-3-031-08223-8_3

Stehlin, J. G., & Payne, W. B. (2022). Mesoscale infrastructures and uneven development: Bicycle sharing systems in the United States as "already splintered" urbanism. Annals of the American Association of Geographers, 112(4), 1065–1083.

Stilo, L., Segura-Velandia, D., Lugo, H., Conway, P. P., & West, A. A. (2021). Electric bicycles, next generation low carbon transport systems: A survey. Transportation Research Interdisciplinary Perspectives, 10, 100347.

Teixeira, P. G., Lebtag, B. G. A., dos Santos, R. P., Fernandes, J., Mohsin, A., Kassab, M., & Graciano-Neto, V. V. (2020). Constituent system design: A software architecture approach. ICSA Companion (pp. 218–225). Salvador, Brazil.

United Nations. (2018). "68% of the World Population Projected to Live in Urban Areas by 2050, Says UN". Available at: http://tiny.cc/ow6ysz

Zaparolli, D. (2022). O futuro da mobilidade com carros autônomos (The future of mobility with autonomous cars). Revista FAPESP. Edition 315. Available at: https://revistapesquisa.fapesp.br/o-futuro-da-mobilidade-com-carros-autonomos/

Zhang, H., Shaheen, S. A., & Chen, X. (2014). Bicycle evolution in China: From the 1900s to the present, International Journal of Sustainable Transportation, 8:5, 317–335, https://doi.org/10.1080/15568318.2012.699999.

8 Building Healthy Cities with Smart Technologies*

8.1 INTRODUCTION

The growing burden on healthcare services worldwide can be primarily attributed to unhealthy lifestyles and living environments (Lee & Nakamura, 2021). According to the World Health Organization (WHO), true health encompasses not just the absence of disease or infirmity, but a state of physical, mental, and social well-being (WHO, 1946; Kuhn & Rieger, 2017). To address this, the WHO has launched the Healthy Cities initiative, which aims to create a network of cities that promote holistic approaches to improving urban health, with a particular focus on the physical and social environments. With an estimated 68% of the world's population projected to be living in urban areas by 2050 (WHO, 2022), the need for actions to improve the health of citizens in these areas has never been greater.

The idea of healthy cities is not a novel one, with early references dating back to the mid-1980s (Duhl, 1986). Duhl and later researchers such as Goldstein and Kickbusch (1996) have emphasized that a healthy city is one that continually strives to improve the physical and social environment for its population, enabling them to reach their full potential, regardless of their economic status. In essence, a healthy city must ensure health for all, which entails reducing disparities in access to health services, emphasizing disease prevention, fostering intersectoral cooperation, reducing environmental hazards, promoting community engagement, prioritizing primary healthcare, educating patients on their health, and participating in national and international efforts to prevent the spread of disease (Duhl, 1986; Goldstein & Kickbusch, 1996).

In recent years, the advent of cutting-edge technologies such as the Internet, cloud computing, medical devices, the Internet of Things (IoT), robotics, and Artificial Intelligence (AI) has opened up a world of possibilities for creating innovative solutions that can bring the concept of healthy cities to fruition. These technologies can enable the development of smart cities that promote healthy citizens and enhance the quality of life.

Creating a truly healthy city requires the collaboration and integration of a wide range of individuals, areas, organizations, specialties, and technologies, all working together to enhance the health of citizens. Figure 8.1 illustrates the various components that make up the ecosystem of a healthy city, depicted as layers. Stakeholders

---

* Portions of this chapter were contributed by Lina Garcés. Available at: https://orcid.org/0000-0002-4990-6562

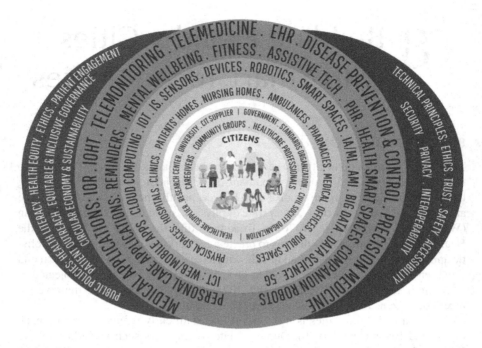

FIGURE 8.1 Overview of a healthy city ecosystem.

(citizens), that is, people and organizations involved in a healthy city, are at the center of the concerns (and of Figure 8.1). Different urban environments, such as hospitals and patients' homes, are also illustrated.

Around this, we have existing technologies that could support the creation of a healthy city, such as IoT and the Cloud. Over that core, a range of possibilities to develop custom applications for citizens and health professionals are also illustrated. Around this entire infrastructure, policies and technical principles for the planning and construction of a healthy city are also there. Details of each layer are presented in the remainder of this chapter. In particular:

- Section 8.2 explains the stakeholders, i.e., people and organizations involved in a healthy city.
- Section 8.3 details the different urban environments that could be part of a healthy city.
- Section 8.4 lists existing technologies that could support the creation of a healthy city.
- Sections 8.5 and 8.6 present a discussion on the range of possibilities for customer applications, for citizens and health professionals, that can be offered in a healthy city.
- Sections 8.7 and 8.8 discuss the policies and technical principles that must be considered during the planning and construction of a healthy city.

8.2 STAKEHOLDERS

A stakeholder in a healthy city is anyone or any entity that is impacted by the operation of the healthcare system and/or has a direct or indirect influence on the quality of healthcare services provided. Within the context of healthcare ecosystems, the following categories of stakeholders can be identified, as shown in the center of Figure 8.1 (Huch, 2010):

- **Primary stakeholders:** The citizens and their families should be the primary focus of any healthcare service within a healthy city ecosystem. These include individuals who utilize, receive or are clients of healthcare services, such as senior citizens, people with disabilities, or family members and relatives who act as private caregivers.
- **Secondary stakeholders:** These are professional users of healthcare systems who have a business-to-consumer (B2C) relationship with the primary stakeholders, as they provide healthcare services to patients. Additionally, they have a business-to-business (B2B) relationship with tertiary stakeholders, as they purchase healthcare systems from suppliers.
- **Tertiary stakeholders:** These are suppliers of healthcare systems, such as research organizations, companies that provide telemedicine or telecare services (e.g., Bosch, Philips, Tunstall), and providers of IT infrastructure (e.g., networks and databases or small and medium-sized enterprises). This category also includes civil engineers, urban architects, and other engineers in areas such as mechanics, electronics, computing, robotics, control, and automation, who play an important role in creating physical environments (e.g., houses, buildings, roads, parks, hospitals, clinics) that promote inclusion, sustainability, and healthy habits.
- **Quaternary stakeholders:** These are individuals or groups that support healthcare systems, such as policymakers, insurance companies, employers, public administrations, standardization organizations, and civil society organizations.

8.3 PHYSICAL SPACES

The second layer of a healthy city's ecosystem comprises all the physical spaces where citizens can access healthcare services. These include healthcare facilities such as hospitals, emergency rooms, clinics, physicians' offices, pharmacies, and ambulances, as well as public spaces that are designed to promote well-being, such as urban squares, nature parks, bike lanes, and other venues for physical activity.

8.4 THE SPECTRUM OF HEALTHCARE-RELATED TECHNOLOGIES

8.4.1 WEB AND MOBILE

Web and mobile technologies deliver solutions to final users more easily and flexibly through the Internet connection. Users can access applications' functionalities, anytime, anywhere. Web and mobile solutions can facilitate smart health in several ways:

- **Access to healthcare:** Web and mobile solutions provide easy access to healthcare services, making it possible for individuals to receive medical care from anywhere, at any time.
- **Remote monitoring:** With the help of mobile and web-based solutions, patients can monitor their health remotely, which can improve patient outcomes and reduce costs.
- **Electronic health records (EHRs):** Web and mobile solutions enable the creation and sharing of EHRs, which can improve the quality of care by providing healthcare providers with a comprehensive view of a patient's health history.
- **Medical communications:** Web and mobile solutions facilitate communication between patients and healthcare providers, making it possible for patients to receive medical advice and consultations remotely.
- **Telemedicine:** Web and mobile solutions can be used to provide telemedicine services, which allow patients to receive medical care remotely, via videoconferencing, for example.
- **Personalized medicine:** Web and mobile solutions can be used to collect and analyze data from patients, which can be used to create personalized treatment plans and improve overall patient outcomes.
- **Patient engagement:** Web and mobile solutions can be used to engage patients in their own care, which can improve patient outcomes, patient satisfaction, and patient adherence to treatment plans.

8.4.2 Information Systems

An Information System (IS) is a sociotechnical system that has as its main components data, communication networks, hardware, software, and people. In an IS, all its components are coordinated to achieve organizational goals through data treatment and valuable information generation (Piccoli & Pigni, 2018; Laudon & Laudon, 2021). Health IS can be used to collect, store, and analyze large amounts of data from various sources, such as EHRs, medical devices, and patient-generated data, which can be used to identify patterns, trends, and insights that can inform decision-making and improve care.

8.4.3 Sensors and Devices

Sensors are specialized hardware that detects environmental signals and converts them into data that can be used by computer systems. This data can include things such as body temperature, heart rate, insulin levels, sleep quality, and air quality, among others. Devices, which are made up of both hardware and software, utilize sensors to process data and generate information that can be shared with other systems or displayed to users. Examples of devices include heart monitors, infusion pumps, and insulin pumps (Al-Kahtani et al., 2022).

8.4.4 IoT

The IoT plays a significant role in smart cities (as discussed in Chapter 2), particularly, in healthcare systems by providing the ability to connect and communicate with various medical devices and sensors that can collect, store, and transmit data. Some examples of the role of IoT in healthcare include remote monitoring, wearable devices, supply chain management for healthcare products, etc.

The use of IoT to provide health services is known as Healthcare IoT (HIoT). HIoT can be broadly categorized into Personal HIoT (PHIoT) and Clinical IoT (CIoT), both technologies are explained in Section 8.5.

8.4.5 Cloud Computing

Cloud computing has been a trend in ICT (Information and Communication Technologies) providing different services to create flexible, available, scalable, and fault-tolerant applications. Cloud is an allusive term to describe a full infrastructure of remote servers in which you can deploy your files and information. Remarkable examples of it are Dropbox and Google Drive, which allow to remotely access your information from anywhere. The same principle could be used in private clouds exclusively dedicated to health applications and storing Personal Health Records, as explained in Section 8.5. Cloud computing offers the backbone to support the availability of other solutions, such as web, mobile, and IoT systems for healthcare. Services provided through the cloud are:

- **IaaS – Infrastructure-as-a-Service:** Making it possible to use hardware infrastructure (e.g., machines configured with different operating systems) as a service. This means that a cloud provider offers virtualized computing resources over the Internet. These resources typically include computing power, storage, and networking infrastructure, which customers can rent on a pay-as-you-go basis.
- **PaaS – Platform-as-a-Service:** Offering the possibility of using a different application, database servers, as well as virtual machines, as remote services that can be invoked by other health applications, for instance storing information;
- **SaaS – Software-as-a-Service:** Providing services such as load-balancing, distributed access control, observability, and monitoring, among others (Sunyaev, 2020).

These services can be used in a wide range of scenarios in the healthcare domain including creating and managing custom solutions (e.g., medical imaging systems, telemedicine platforms, clinical decision support) that can be tailored to the specific needs of healthcare providers and patients.

8.4.6 Robots

Robots are technologies conceived to support humans in repetitive or dangerous situations. The development of robots requires multidisciplinary teams, in technical areas such as design, electronics, mechanics, computer sciences, materials engineering, mathematics, automation, control, and software engineering, as well in domain specialties in which robotics is applied as healthcare, agriculture, industry, among others (Guntur et al., 2019). Robots are used in healthcare in a variety of ways. Some examples include:

- **Surgical robots:** These robots assist surgeons in performing complex procedures, such as minimally invasive surgeries. They offer increased precision and control, as well as a reduced risk of complications.
- **Rehabilitation robots:** These robots are used to help patients recover from injuries or surgeries. They can help with exercises, balance training, and even assist in walking.
- **Telepresence robots:** These robots allow healthcare providers to remotely interact with patients. They can be used for telemedicine consultations, remote patient monitoring, and even for socialization with patients in isolation.
- **Automated medication dispensers:** These robots are used to dispense medication to patients in a hospital setting. They can help reduce errors in medication administration and improve patient safety.
- **Cleaning robots:** These robots are used to disinfect and clean hospitals, clinics, and other healthcare facilities, they can move around autonomously and they can detect the areas that need more cleaning and attention.

8.4.7 AI/ML

AI and Machine Learning (ML) are areas of computer science that aims at the creation of solutions inspired by human intelligence, including mimicking cognitive functions such as perception, learning, and problem-solving.

In some scenarios, the application of AI/ML in healthcare could improve patients' diagnostics, prevention, and treatment, increasing cost efficiency and equality, and equality in health services (Sunarti et al., 2021). AI/ML has the potential to revolutionize the healthcare industry by improving patient outcomes and reducing costs. Some examples of the use of AI/ML in healthcare include:

- **Medical diagnosis:** AI-powered systems can assist doctors in diagnosing diseases by analyzing medical images and identifying patterns that may be missed by human eyes.
- **Predictive analytics:** AI-powered systems can analyze patient data to predict the likelihood of developing certain conditions, such as heart disease or cancer, and help doctors make better treatment decisions.
- **Personalized medicine:** AI and ML can be used to analyze patient data to identify personalized treatment options that are most likely to be effective.
- **Drug discovery:** AI and ML can be used to analyze large datasets of chemical compounds to identify new drug candidates.
- **Medical imaging:** AI can help radiologists and other medical professionals to analyze medical images, such as CT and MRI scans, to identify potential issues, such as tumors or other abnormalities.

- **Medical chatbots:** AI-powered chatbots can be used to assist patients in finding the right care and to answer questions they may have about their health.

8.4.8 AMBIENT INTELLIGENCE (AMI)

Ambient Intelligence (AmI) is the integration of intelligence into everyday environments, making them responsive to human needs. It is related to fields such as pervasive computing, ubiquitous computing, and AI. AmI systems possess characteristics such as being invisible or transparent, mobile, context-aware, anticipatory, adaptive, sensitive, ubiquitous, and responsive. They are also equipped with sensors and wireless communication interfaces to scan the environment and exchange information with other devices. Additionally, AmI incorporates AI, including ML, agent-based software, and robotics. AmI can be used to create intelligent hospitals that can optimize patient flow, reduce errors and improve the overall quality of care. It can be used to predict and respond to critical situations, such as heart attacks or sudden falls, in a timely manner.

8.5 PERSONAL HEALTH APPLICATIONS

The integration of various technologies, including the web, mobile devices, sensors, the IoT, and cloud computing, has led to the development of innovative and significant personal health-related solutions. This section provides a comprehensive look at various applications, systems, and technologies that enable autonomous patient healthcare.

8.5.1 HEALTH REMINDERS APPLICATIONS

Health reminder applications are typically designed to run on mobile devices, and allow users to set reminders for various health-related activities, such as taking medication at the correct time, monitoring diet, scheduling appointments, tracking symptoms, and reminding them to engage in physical activity or follow specific pharmacological treatments. These types of apps are widely available and can be found in various sources such as Creveling and Goldman (2021).

8.5.2 MENTAL WELL-BEING APPLICATIONS

Digital technologies, including social networks and mobile applications, can be utilized to provide mental health interventions. These interventions have been shown to be effective in addressing a variety of mental health issues, such as depression, stress, anxiety, and smoking cessation (Harrison et al., 2011). Examples of mental well-being applications include:

- **Headspace:** A mindfulness and meditation app that offers guided meditations, sleep soundscapes, and mindfulness exercises.
- **Moodfit:** An app that provides personalized mental fitness training to help users improve their mood and reduce stress.

- **Talkspace:** A therapy app that connects users with licensed therapists for online counseling and therapy sessions.
- **Pacifica:** An app that offers daily tools for managing stress, anxiety, and depression, including mood tracking, guided meditations, and cognitive behavioral therapy exercises.
- **Happify:** An app that offers science-based activities and games to improve emotional well-being and reduce stress and anxiety.
- **Calm:** An app that offers guided meditations, sleep stories, and music to help users relax and improve their mental well-being.

A comprehensive review of apps for mental health support can be found in Giota and Kleftaras (2014).

8.5.3 Fitness Applications

Fitness apps are mobile applications that assist users in tracking and monitoring their physical activity and exercise. They cover a range of sports, such as biking, hiking, running, gym, and dance, and can be used for solo or group workouts. These apps also help users to monitor their diet and nutrition, as well as track body metrics such as weight and BMI. Examples of fitness apps include MyFitnessPal, Nike Training Club, Fitbit, 7 Minute Workout, JEFIT, StrongLifts 5x5, and Couch to 5K. These apps provide features such as workout tracking, personalized exercise plans, progress tracking, and community support. Some apps also integrate with wearable fitness devices such as smartwatches to provide more accurate data tracking. More examples of fitness apps can be found in a study conducted by Hall et al. 2022.

8.5.4 Assistive Technology

Ambient-Assisted Living (AAL) refers to the use of technology to support older adults and people with disabilities in their daily lives, allowing them to live independently for as long as possible. This can include a wide range of assistive technologies such as sensors, wearables, and home automation systems, which can help with tasks such as monitoring health, providing reminders, and enabling remote communication with caregivers or family members. The AAL concept emerged in the 1990s, and since the middle of the 2000s, it has received more attention. AAL is a relatively new field and has become an increasingly important, multi-disciplinary research topic for both the medical and the technological research communities. Examples of AAL include voice-controlled devices that can be used to control lighting and appliances and to make phone calls or send text messages. Another example is virtual reality systems that can be used for physical and cognitive therapy, as well as for socialization and entertainment.

8.5.5 Personal Health Records (PHR)

PHR is an electronic record of health-related information on an individual that can be shared across different healthcare settings. It includes information such as demographics, medical history, medications, allergies, lab results, and other relevant health data. The goal of PHR is to provide individuals with access to their health information and empower them to take a more active role in managing their health. Generally speaking, PHR serves a variety of purposes, including:

a. Information Collection, such as past medical history, family history, allergies, medications, and logs for health-related activities such as mood and sleep tracking, and glucose monitoring.
b. Information sharing between patients and healthcare providers.
c. Support for self-management, with features such as action plans for managing chronic diseases.
d. Facilitation of information exchange, for tasks such as appointment scheduling and managing medications (Kaelber & Pan, 2008).

Examples of PHR systems include Microsoft HealthVault, Google Health, and MyFitnessPal. This information can be stored in blockchain, as discussed in Chapter 6.

8.5.6 PERSONAL HEALTH IoT (PHIoT)

PHIoT, or Personal Health IoT, encompasses a variety of devices and applications that are used for self-monitoring purposes. These can include activity and heart-rate trackers, smart clothes, and smartwatches. It is important to note that the majority of these devices are not currently regulated and are intended for use by individuals without physician involvement or guidance. Examples of PHIoT include mobile apps that are connected to smartwatches, such as the Apple Watch or Samsung Health, which allow for the monitoring of health parameters such as heart rate, oxygen saturation, sleep quality, and activity.

8.5.7 COMPANION ROBOTS

Companion robots, such as butler robots, provide companionship and assistance to individuals who spend a significant amount of time alone, such as the elderly, disabled individuals, or children. These robots offer a range of capabilities, from simple interactions like those of Alexa to comprehensive assistance with daily living activities (Lamers & Verbeek, 2011; Odekerken-Schröder et al., 2020). Examples of companion robots include:

- Pepper, a humanoid robot developed by SoftBank Robotics that can recognize and respond to human emotions.
- Jibo, a social robot designed for the home that can assist with tasks and entertain with interactive stories and games.
- Nannybot, a robot designed to assist with child care and monitoring, including video and audio monitoring, remote control of household appliances, and providing educational games and activities for children.

- Elderly companion robots, such as PARO, a therapeutic robot that is designed to provide comfort and companionship for older adults living in assisted living or long-term care facilities.
- Kompai, a robot designed to assist with tasks such as reminding patients to take their medication and monitoring vital signs.

8.6 APPLICATIONS SUPPORTING HEALTHCARE PROFESSIONALS

This section highlights various technologies and applications that aid healthcare professionals in their work.

8.6.1 ELECTRONIC HEALTH (E-HEALTH)

Electronic health (e-health) refers to the use of technology and digital tools to manage, store, and access health information, and to support health-related activities such as communication, diagnosis, treatment, and care.

The history of e-health, or electronic health, dates back to the 1960s when the first electronic medical records (EMRs) were developed. In the 1970s and 1980s, the use of computers in healthcare increased with the development of more sophisticated EMR systems and the use of telemedicine for remote consultations. In the 1990s, the Internet began to be used for healthcare purposes, leading to the development of online health information and communication between patients and healthcare professionals. In the 2000s, the use of mobile technology in healthcare, such as smartphones and tablets, led to the development of new e-health applications, such as telemedicine and mHealth apps. Nowadays, e-health includes a wide range of digital technologies and services, from EHRs and telemedicine to mHealth apps and AI, that are used to improve the efficiency, effectiveness, and accessibility of healthcare.

8.6.2 ELECTRONIC HEALTH RECORDS (EHR)

The use of electronic healthcare records (EHRs) has its roots in the 1960s, with the development of the first computerized medical records systems. These early systems were primarily used for administrative tasks, such as billing and scheduling. In the 1980s and 1990s, EHRs began to be adopted by larger healthcare organizations and government agencies, as computer technology became more advanced and more widely available.

The development of the Internet in the late 1990s and early 2000s led to a new generation of EHRs that were accessible remotely and could be shared among different healthcare providers. This helped to improve patient care and reduce errors, as healthcare providers could access patient information from multiple sources.

In the 2000s and 2010s, the U.S. government and private sector began to invest heavily in the development and adoption of EHRs, through initiatives such as the HITECH Act and the Meaningful Use program. These efforts have helped to increase the number of healthcare providers using EHRs, and have also led to the development of new standards and regulations for EHRs, such as the HL7 and CDA standards. Recently there has been active research on using blockchain technology as a platform for HER (Kassab et al., 2019).

However, the widespread adoption of EHRs has also faced some challenges, such as concerns about data privacy and security, and the cost and complexity of implementing and maintaining EHR systems. Despite these challenges, EHRs are now considered a standard part of healthcare delivery and are expected to play an increasingly important role in the future of healthcare.

8.6.3 CLINICAL INTERNET OF THINGS (CIoT)

CIoT, or Clinical Internet of Things, refers to technology and devices used in the diagnosis and treatment of patients' health conditions. Examples of CIoT include smart connected inhalers, continuous glucose monitors, and smart heart rate monitors. Additionally, Integrated Operation Rooms (IORs) are also considered a type of CIoT, as they connect all devices within surgical rooms and allow for centralized control and access to surgery information.

In contrast to PHIoT, or Personal Health Internet of Things, CIoT devices send their output data to health ISs for use by clinical professionals, rather than directly to the consumer's storage device. Additionally, CIoT data is intended to be shared with patients and possibly other systems such as EHRs.

Recent trends have also aimed to integrate PHIoT and CIoT solutions in order to extend clinical services to patients' homes. The Healthcare Supportive Home (HSH) is an example of this, as it uses technology such as HIoT, AmI, eHealth, sensor networks, assistance robots, advanced human–machine interfaces, microelectronics, and web and mobile apps to provide patients with autonomy and support in their daily lives. An example of this is an HSH system for remotely monitoring patients with diabetes.

8.6.4 DISEASE PREVENTION AND CONTROL

Mobile apps have been developed by organizations focused on disease control and prevention to distribute crucial information about illnesses, including symptoms, causes, and actions to be taken by the public during critical situations. These apps allow citizens to access updated information about preventing common diseases and measures to avoid viral or bacterial infections. For example, there has been a flood of mobile applications such as contact tracing apps in the middle of the battle with COVID-19 (Kassab et al., 2021).

8.6.5 PRECISION MEDICINE

Precision medicine, also referred to as personalized medicine, is an approach to disease prevention and treatment that takes into account an individual's unique genetic makeup, environment, and lifestyle. The goal of this initiative is to provide the most appropriate treatment to the right patient at the right time. This is achieved by analyzing a patient's genotypic data, such as DNA variations, and phenotypic data, which includes traditional clinical biomarkers, gene expression, and metabolite profiles (Ginsburg & Haga, 2019).

8.7   PUBLIC POLICIES

The successful implementation of healthy cities involves a variety of strategies and policies that guide health-related organizations to achieve:

- Health literacy, which equitably empowers citizens to access, understand, and use information and services to make informed health-related decisions for themselves and others (CDC, 2022a).
- Health equity, which ensures that every person has the opportunity to reach their full health potential without being disadvantaged by social or other determinants of circumstance (CDC, 2022b).
- Patient engagement, which enables individuals to make informed choices about their care options and effectively utilize health-related resources, thereby promoting the sustainability of health systems worldwide (WHO, 2016).
- Patient outreach, which involves any form of communication with patients to keep them involved in their care, improve adherence to care plans, and prompt them to take action (RELATIENT, 2022).
- Equitable and inclusive governance, which includes policies that reduce or eliminate social disparities in healthcare access, specifically related to class, gender, ethnicity, geography, and various forms of discrimination or social exclusion (WHO, 2014).
- Circular economy and sustainability policies, provide a path to sustainable growth, good health, and decent jobs while protecting the environment and natural resources. A circular economy aims to renew, reuse, and share technological components and materials, thereby avoiding risks of adverse and unintended health effects, particularly in processes involving hazardous materials (Sjödin, 2006; WHO, 2018).

8.8   TECHNICAL PRINCIPLES

Creating a healthy city requires a diverse range of strategies and policies to guide health-related organizations in achieving certain goals. Along with these goals, the implementation of new technologies in a healthy city also poses several challenges that must be addressed by engineers, scientists, and tech providers. These challenges include:

- **Ethics:** New technologies must be developed with an ethical basis to ensure they are making decisions that benefit patients' health. Scientists must consider the sociotechnical system in which the technology is situated and address ethical issues related to decision-making and patient data treatment (Shaw & Donia, 2020).
- **Trust:** Patients must trust in healthcare policies, providers, technologies, and other initiatives to improve their quality of life.
- **Safety:** The healthy city ecosystem must be coordinated to ensure citizens' safety in all their health-related activities, including economic, environmental, physical, and mental safety.

- **Privacy:** A healthy city generates a large amount of citizen information that must be kept confidential to preserve their dignity and protect their human rights. The city must follow national and regional data privacy regulations.
- **Security:** The healthy city infrastructure, data integrity, and citizens' privacy must be protected from security threats such as invasions and data leaks. Security-by-design and risk-based approaches must be used to prevent these issues.
- **Interoperability:** The entire ecosystem must be interconnected to allow all stakeholders to understand and make decisions based on the information generated in a healthy city.
- **Accessibility:** Technologies in a healthy city must be designed to be accessible to all citizens, regardless of their health conditions or disabilities.

8.9 FINAL REMARKS

This chapter delves into the concept of smart cities as a means to enhance the quality of life for citizens through the collaboration and cooperation of various fields and stakeholders. The idea of healthy cities presents new and exciting possibilities for creating an innovative healthcare ecosystem for citizens, communities, professionals, organizations, and governments. The vast array of opportunities and cutting-edge technologies can aid in the creation of products, services, and environments that improve the population's well-being. However, realizing these opportunities requires an understanding of the current barriers and challenges, particularly those related to public policies and technological principles in healthy cities ecosystems, as outlined by Goldstein and Kickbusch (1996) including (i) increasing citizens' awareness of their health and teaching disease prevention strategies; (ii) ensuring access to health services for all citizens; (iii) making public health a priority on the social and political agenda and contributing to the development of healthy municipal policies; (iv) promoting cooperation among health-related organizations, institutions, departments, and sectors; and (v) encouraging community participation to support citizens' well-being. The next chapter covers agribusiness regarding smart cities.

REFERENCECS

AALIANCE. (2010). Ambient Assisted Living Roadmap, v. 6 of Ambient Intelligence and Smart Environments. IOS Press, Amsterdam, p. 136

Al-Kahtani, M. S., Khan, F., & Taekeun, W. (2022). Application of internet of things and sensors in healthcare. Sensors, 22(15), 5738.

Andreu-Perez, J., Poon, C. C. Y., Merrifield, R. D., Wong, S. T. C., & Yang, G. -Z. (2015). "Big data for health," in IEEE Journal of Biomedical and Health Informatics, 19(4), 1193–1208, https://doi.org/10.1109/JBHI.2015.2450362.

Arenas, C., & Garcés, L. (2021). Integrated operating room: A systematic mapping review. Journal of Health Informatics, 2020, 412–419. Available at: https://jhi.sbis.org.br/index.php/jhi-sbis/article/view/847/450. Last access October 15, 2022.

Buchmayr, M. M., & Kurschl, W. W. (2011). A survey on situation-aware ambient intelligence systems. Journal of Ambient Intelligence and Humanized Computing, 2(3), 175–183.

Centers of Disease Control and Prevention (CDC). (2022a). Health Equity. National Center for Chronic Disease Prevention and Health Promotion (NCCDPHP). Available at: https://www.cdc.gov/chronicdisease/healthequity/index.htm. Last Access: October 15, 2022.

Centers of Disease Control and Prevention (CDC). (2022b). What is Health Literacy?. Available at: https://www.cdc.gov/healthliteracy/learn/index.html. Last Access: October 15, 2022.

Constantine, R. (2013). Electronic health records. In: E. M. Sullivan, D. V. Darwin Brown (eds). Physician Assistant: A Guide to Clinical Practice: Expert Consult, 5th ed., Chapter 17, Elsevier, Amsterdam, pp. 221–225.

Cook, D. J., Augusto, J. C., & Jakkula, V. R. (2009). Ambient intelligence: Technologies, applications, and opportunities. Pervasive and Mobile Computing, 5(4), 277–298.

Creveling, M., & Goldman, E. (2021). The 16 Best Health and Wellness Apps of 2021, According to Experts. Prevention. Available at: https://www.prevention.com/health/sleep-energy/g24736063/best-health-apps/. Last Access: October 15, 2022.

Dave, V., & Joshi, N. (2019). "Fog computing enabled Ambient Assisted Healthcare systems," 2019 IEEE International Conference on Distributed Computing, VLSI, Electrical Circuits and Robotics (DISCOVER), pp. 1–7, https://doi.org/10.1109/DISCOVER47552.2019.9007916.

Duhl, L. J. (1986). The healthy city: Its function and its future, *Health Promotion International*, 1(1), 55–60, https://doi.org/10.1093/heapro/1.1.55

Eysenbach, G. (2001). What is e-health? Journal of Medical Internet Research (JMIR), 3(2), e20.

Food and Drug Administration (FDA). (2018). Precision Medicine. Available at: https://www.fda.gov/medical-devices/in-vitro-diagnostics/precision-medicine. Last Access: October 15, 2022.

Garcés, L., Ampatzoglou, A., Avgeriou, P., & Nakagawa, E. Y. (2015). "A Reference Architecture for Healthcare Supportive Home Systems," 2015 IEEE 28th International Symposium on Computer-Based Medical Systems, 2015, pp. 358–359, https://doi.org/10.1109/CBMS.2015.39.

Garcés, L., Zanin Vicente, I., & Nakagawa, E. Y. (2019). "Software Architecture for Health Care Supportive Home Systems to Assist Patients with Diabetes Mellitus," 2019 IEEE 32nd International Symposium on Computer-Based Medical Systems (CBMS), 2019, pp. 249–252, https://doi.org/10.1109/CBMS.2019.00060.

Giffinger, R., Fertner, C., Kramar, H., Kalasek, R., Pichler-Milanovic, N., & Meijers, E. (2007). Smart Cities-Ranking of European Medium-Sized Cities. Technical report, Vienna University of Technology.

Ginsburg, G. S., & Haga, S. B. (2019). Foundations and Application of Precision Medicine. Emery and Rimoin's Principles and Practice of Medical Genetics and Genomics (Seventh Edition), 21–45. https://doi.org/10.1016/B978-0-12-812537-3.00002-0

Giota, K., & Kleftaras, G. (2014). Mental health apps: Innovations, risks and ethical considerations. E-Health Telecommunication Systems and Networks, 3, 19–23. https://doi.org/10.4236/etsn.2014.33003.

Goldstein, G., & Kickbusch I. (1996). A healthy city is a better city. World Health. 49th Year. no. 1. January-February. pp. 4–6. https://apps.who.int/iris/bitstream/handle/10665/330422/WH-1996-Jan-Feb-p4-6-eng.pdf?sequence=1. Last Access: October 15, 2022.

Guntur, S. R., Gorrepati, R. R., & Dirisala, V. R. (2019). Chapter 12 – Robotics in Healthcare: An Internet of Medical Robotic Things (IoMRT) Perspective, in N. Dey, S. Borra, A. S. Ashour, F. Shi (eds). Machine Learning in Bio-Signal Analysis and Diagnostic Imaging, Academic Press, pp. 293–318, https://doi.org/10.1016/B978-0-12-816086-2.00012-6.

Habibzadeh, H., Dinesh, K., Rajabi Shishvan, O., Boggio-Dandry, A., Sharma, G., & Soyata, T. (2020). "A survey of healthcare internet of things (HIoT): A clinical perspective," IEEE Internet of Things Journal, 7(1), 53–71, https://doi.org/10.1109/JIOT.2019.2946359.

Hall, A., & Williams, V.. Best fitness apps of 2022. Forbes. Available at: https://www.forbes.com/health/body/best-fitness-apps/#summary_best_fitness_apps_section. Last Access: October 15, 2022.

Harrison, V., Proudfoot, J., Wee, P.P., Parker, G., Pavlovic, D. H., & Manicavasagar, V. (2011). Mobile Mental health: Review of the emerging field and proof of concept study. Journal of Mental Health, 20, pp. 09–524. http://dx.doi.org/10.3109/09638237.2011.608746

Home Care Management Associates. (1998). Home Tele-Health Systems: A Guide for Home CareProviders. Technical report, Home Care Management Associates Ltd., Springfield, PA.

Huch, M. (2010). D2: Identification and characterisation of the main stakeholder groups for "ICT for Ageing" solutions. Version 5. BRAID project. p. 46. http://www.supras.biz/pdf/Huch-M_2010_StakeholderAnalysis-ICT-Aging.pdf. Last Access: October 15, 2022.

Institute of Medicine. (1996). Telemedicine: A Guide to Assessing Telecommunications for Health Care. Washington, DC: The National Academies Press, 288 p.

Kaelber, D., & Pan, E. (2008). The Value of Personal Health Record (PHR) Systems. In AMIA Annual Symposium Proceedings, Online, pp. 343–347.

Kassab, M., DeFranco, J., Malas, T., Neto, V. V. G., & Destefanis, G. (2019). Blockchain: A panacea for electronic health records?. In 2019 IEEE/ACM 1st International Workshop on Software Engineering for Healthcare (SEH) (pp. 21–24).

Kassab, M. H., Neto, V. V. G., Destefanis, G., & Malas, T. (2021). Could blockchain help with COVID-19 crisis?. IT Professional, 23(4), 44–50.

Kon, F., & Santana, E. F. Z. (2016). Cidades inteligentes: Conceitos, plataformas e desafios. Jornadas De Atualização Em Informática, *17*. Available at: https://sol.sbc.org.br/livros/index.php/sbc/catalog/download/6/6/17-1?inline=o. Last Access: October 15, 2022.

Kuhn, S., & Rieger, U. M. (2017). Health is a state of complete physical, mental and social well-being and not merely absence of disease or infirmity. American Society for Metabolic and Bariatric Surgery, 13(5), 887. https://doi.org/10.1016/j.soard.2017.01.046

Lamers, M. H., & Verbeek, F. J. (2011). Human-Robot Personal Relationships: Third International Conference, HRPR 2010, Leiden, The Netherlands, June 23–24, 2010, Revised Selected Papers.Springer Science & Business Media, Heidelberg, p. 13.

Laudon, K. C., & Laudon, J. P. (2021). Management Information Systems: Managing the Digital Firm. Pearson Education Limited, London. 17th edition.

Lee, A., & Nakamura, K. (2021). Engaging diverse community groups to promote population health through healthy City approach: Analysis of successful cases in Western pacific region. International Journal of Environmental Research and Public Health, 18(12), 6617. https://doi.org/10.3390/ijerph18126617

Maheu, M., Whitten, P., & Allen, A. (2001). E-health, telehealth, and telemedicine: A guide to startup and success. Jossey-Bass, 1–400.

Meystre, S. (2005). The current state of telemonitoring: A comment on the literature. Telemedicine Journal and e-Health: the Official Journal of the American Telemedicine Association, 11(1), 63–69.

Moore, M. (1999). The evolution of telemedicine. Future Generation of Computer Systems, 15(2), 245–254.

Nehmer, J., Becker, M., Karshmer, A., & Lamm, R. (2006). Living assistance systems: an ambient intelligence approach. In 28th international conference on Software engineering, Shanghai, China, ACM, pp. 43–50.

Odekerken-Schröder, G., Mele, C., Russo-Spena, T., Mahr, D., & Ruggiero, A. (2020), "Mitigating loneliness with companion robots in the COVID-19 pandemic And beyond: An integrative framework and research agenda", Journal of Service Management, 31(6), 1149–1162. https://doi.org/10.1108/JOSM-05-2020-0148

Piccoli, G., & Pigni, F. (2018). Information Systems for Managers: with Cases (Edition 4.0 ed.). Prospect Press, Burlington, VT. p. 28. Retrieved 25 November 2018.

Pieper, M., Antona, M., & Cortes, U. (2011). Ambient assisted living. Ercim News, 87, 18–19, https://ercim-news.ercim.eu/images/stories/EN87/EN87-web.pdf

Relatient.com. (2022). Effective patient outreach: how to, do's and don'ts. Available at: https://www.relatient.com/patient-outreach/#:~:text=What%20Is%20Patient%20 Outreach%3F,prompt%20them%20to%20take%20action.

Rodríguez, L. M. G. (2018). A reference architecture of healthcare supportive home systems from a systems-of-systems perspective. Doctoral Thesis, Instituto de Ciências Matemáticas e de Computação, University of São Paulo, São Carlos. https://doi. org/10.11606/T.55.2018.tde-16102018-111654. Retrieved October 17, 2022, from www. teses.usp.br

Shaw, J. A., & Donia, J. (2020). The sociotechnical ethics of digital health: A critique and extension of approaches from bioethics. Frontiers in Digital Health. https://doi. org/10.3389/fdgth.2021.725088

Sjödin, A. (2006). High tech trash: Digital devices, hidden toxins, and human health. Environmental Health Perspectives, 114(8), A500. https://www.ncbi.nlm.nih.gov/pmc/ articles/PMC1551981/

Sunarti, S., Fadzlul Rahman, F., Naufal, M., Risky, M., Febriyanto, K., & Masnina, R. (2021). Artificial intelligence in healthcare: Opportunities and risk for future. Gaceta Sanitaria, 35, S67–S70. https://doi.org/10.1016/j.gaceta.2020.12.019

Sunyaev, A. (2020). Cloud computing. In Internet Computing. Springer, Cham. https://doi. org/10.1007/978-3-030-34957-8_7

Tang, P., Ash, J., Bates, D., Overhage, J., & Sands, D. Z. (2006). Personal health records: Denitions, benets, and strategies for overcoming barriers to adoption. Journal of the American Medical Informatics Association (JAMIA), 13(2), 121–126.

Tazari, S., Valero, A. F., Dommarco, R., Lazaro Ramos, J. P., & Furfari, F. (2010). PERSONA PERceptive Spaces prOmoting iNdependent Aging. nal reference architecture model for aal and recommendations for future activities on the Open AAL Platform. Online. http://www.aal-persona.org. Accessed in June 15, 2021. D3.1.3, Fraunhofer IGD, ITACA UPV, and CNR-ISTI.

Weber, W., Rabaey, J., & Aarts, E. eds. (2005). Ambient Intelligence. 14th ed. Springer, Berlin, 374 p.

World Health Organization (WHO). (1946). Constitution. Available at: https://www.who.int/ about/governance/constitution. Last Access: October 15, 2022.

World Health Organization (WHO). (2014). Governance for health equity. Taking forward the equity values and goals of Health 2020 in the WHO European Region. Available at: https://www.euro.who.int/__data/assets/pdf_file/0020/235712/e96954.pdf. Last Access: October 15, 2022.

World Health Organization (WHO). (2016). Patient Engagement. Technical Series on Safer Primary Care. TechReport. ISBN 978-92-4-151162-9. Available at: https://apps.who.int/ iris/bitstream/handle/10665/252269/9789241511629-eng.pdf. Last Access: October 15, 2022.

World Health Organization (WHO). (2018). Circular economy and health: Opportunities and risks. ISBN 9789289053341. Available at: https://www.euro.who.int/__data/assets/pdf_ file/0004/374917/Circular-Economy_EN_WHO_web_august-2018.pdf. Last Access: October 15, 2022.

World Health Organization (WHO). (2022). Healthy Cities Network. Available at: https:// apps.who.int/health-topics/urban-health/cities-spotlight/who-healthy-cities-network. Last Access: October 15, 2022.

9 Agribusiness in the Era of Smart Cities
Opportunities and Challenges*

9.1 INTRODUCTION

The sustainability of urban centers is closely linked to the production processes that provide for their needs. The provision of food is a critical aspect, with plant and animal products sourced from rural properties at the beginning of the production chain. In order to ensure that the supply of goods and services to smart cities is continuous and sustainable, it is essential that the production processes are also optimized. To that end, the concepts of smart farms, smart agriculture, and smart livestock have gained attraction as a means of achieving this goal. Through the use of innovative technologies and advanced data management techniques, smart agribusiness aims to optimize food production while reducing environmental impact and promoting long-term sustainability.

Agribusiness refers to the economic activities related to agricultural production, including (i) *vegetable farming* (henceforth, agriculture), (ii) *livestock raising*, and (iii) *poultry farming*, as the main topics of this chapter. The Agribusiness 4.0 includes smart agriculture and smart livestock. Smart livestock farming is a modern approach that utilizes digital technologies to improve productivity, efficiency, and sustainability of farm operations for animal raise. In turn, smart agriculture, also known as precision agriculture or digital agriculture, refers to the use of advanced technologies and data analytics to improve the efficiency, sustainability, and productivity of agricultural practices. This chapter focuses on smart agriculture, also addressing livestock (animals raise) and poultry farming that, despite the similarities, are addressed by knowledge authorities as a separate topic.

9.2 SMART AGRICULTURE IN RESEARCH AND PRACTICE

Smart agriculture encompasses a wide range of techniques aimed at optimizing crop production, resource use, and environmental sustainability. The use of sensors, automation systems, drones, and satellite imagery are among the tools and approaches that fall under the umbrella of smart agriculture. These technologies help farmers

---

* Portions of this chapter contributed from José Maria David, Wagner Arbex, Regina Braga, and Jonas Gomes.

DOI: 10.1201/9781003348542-9

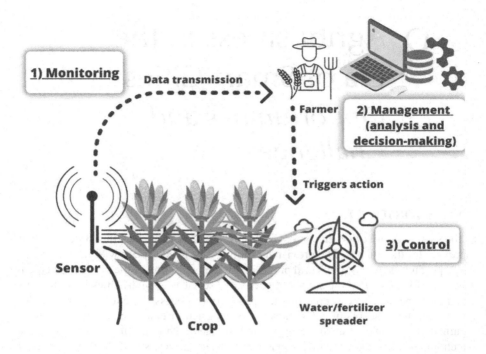

FIGURE 9.1   Smart agriculture main activities and its dynamics.

and agribusinesses make informed decisions, leading to increased efficiency, productivity, and reduced costs and environmental impact.

Smart agriculture also has the potential to improve food security by enabling farmers to grow more food on less land and adapt to changing environmental conditions. The goal is to create a more sustainable and efficient agricultural system that can meet the growing demands of the global population. The Internet of Things (IoT), discussed in Chapter 2 of this book, plays a significant role in smart agriculture, and there is a growing body of research exploring the ways in which it can contribute to the field.

Smart agriculture activities can be broadly categorized into three areas: monitoring, management, and control, as shown in Figure 9.1.

Monitoring is the activity delivered by sensors. Humidity, temperature, plant sizes, and overall production are examples that can be monitored using sensors. Monitoring cultures have the potential to prevent losses and improve overall productivity. Sensors can also be used with animals to measure similar data. The data collected by sensors serve as input for decisions regarding the farm budget and the administration of the entire production, what is known as management, the following step.

Management is concerned with the activities related to supporting a precise analysis and strategic decisions about production. For instance, if a relevant part of the soy production has a size lower than expected, this can be an input parameter for a management system that can adjust/suggest the commercialization prices.

Finally, *control* is related to the actuators and how they support the activities that can be automated or semi-automated, such as (i) watering plants with the right amount of water required by that specific type of crop, (ii) delivering the ideal and empirically proven amount of pesticides that can protect the plant from pests while maintaining low levels of toxicity for the final consumer, or (iii) harvest in an automated way, optimizing the use of plants bodies, avoiding losses and taking care of sustainability, and the natural resources.

The deployment of smart farms has been found to have numerous benefits, such as promoting sustainability and reducing the ecological impact of agricultural production. It can also automate repetitive and continuous work, facilitate fast and precise decision-making, and reduce the need for human presence at monitoring sites. Additionally, many of the technologies used in smart agriculture, such as those based on Arduino, are financially accessible.

There are many case studies of smart agriculture projects that have been implemented around the world. Here are a few examples:

1. **Smart irrigation in California:** In California, a project called the "Smart Irrigation Technology Adoption Program" is helping farmers use sensors and software to optimize irrigation schedules and reduce water use. By using sensors to measure soil moisture levels and weather data to predict future conditions, farmers are able to make more informed decisions about when and how much to water their crops.

2. **Precision agriculture in Australia:** In Australia, a project called the "Precision Agriculture for Development" is using drones and other technologies to collect data on crop health, soil conditions, and other factors that affect crop yields. This data is then used to create customized fertilization and pest control plans for each field, helping farmers to increase yields and reduce the use of resources.

3. **Smart agriculture in Kenya:** In Kenya, a project called the "Smart Agriculture for Improved Livelihoods" is using IoT sensors and other technologies to collect data on soil moisture levels, weather patterns, and crop health. This data is then used to optimize irrigation and fertilization schedules, and to identify potential issues before they become problems. The project has helped to improve crop yields and increase the income of smallholder farmers.

4. **Smart greenhouse in the Netherlands:** A project in the Netherlands is using IoT sensors and machine learning to optimize the temperature, humidity, and light levels in a greenhouse, resulting in improved crop yields and reduced energy use.

5. **Smart irrigation in India:** In India, a project called the "Smart Irrigation System" is using IoT sensors and other technologies to optimize irrigation schedules and reduce water use. The project has helped to improve crop yields and increase the income of smallholder farmers.

6. **Precision agriculture in the United States:** In the United States, a project called the "Precision Agriculture Development Center" is using drones, global positioning system (GPS) mapping, and other technologies to collect

data on soil conditions, crop health, and other factors that affect crop yields. This data is then used to create customized fertilization and pest control plans, helping farmers to increase yields and reduce the use of resources.

7. **Smart agriculture in China:** In China, a project called the "Smart Agriculture Demonstration Park" is using IoT sensors and other technologies to collect data on soil moisture levels, weather patterns, and crop health. This data is then used to optimize irrigation and fertilization schedules, and to identify potential issues before they become problems. The project has helped to increase crop yields and improve the sustainability of agriculture.

8. **Smart irrigation in Brazil:** In Brazil, a project called the "Smart Irrigation System" is using IoT sensors and other technologies to optimize irrigation schedules and reduce water use. The project has helped to improve crop yields and increase the income of smallholder farmers.

9. **Precision agriculture in France:** In France, a project called the "Precision Agriculture for Sustainable Intensification" is using drones and other technologies to collect data on soil conditions, crop health, and other factors that affect crop yields. This data is then used to create customized fertilization and pest control plans, helping farmers to increase yields and reduce the use of resources.

10. **Smart agriculture in South Africa:** In South Africa, a project called the "Smart Agriculture Initiative" is using IoT sensors and other technologies to collect data on soil moisture levels, weather patterns, and crop health. This data is then used to optimize irrigation and fertilization schedules, and to identify potential issues before they become problems. The project has helped to increase crop yields and improve the sustainability of agriculture.

9.3 CHALLENGES FACING SMART AGRICULTURE

While the integration of advanced technologies and data analytics in agriculture holds immense promise for the production of food and management of natural resources, it also presents a range of obstacles that must be overcome. One such challenge is the adaptation of IoT-based systems to different forms of agriculture, such as animal farming, planting, or cultivation, which requires a shift in the way solutions are specified to allow for deployment in different domains. Additionally, there may be opposition to the use of technology in agriculture from those who view it as a threat to traditional practices or the environment, presenting challenges for farmers and agribusinesses seeking to implement smart agriculture solutions.

To address these challenges, there is a need for solutions that are flexible and adaptable to various applications, and for addressing the concerns of those who are wary of the integration of technology in agriculture.

Selecting sensors with the required level of accuracy is another significant challenge in precision agriculture. Some sensors may not deliver adequate data or may not be suitable for use in large-scale operations. Additionally, access to technology and data analytics tools can be restricted in certain regions, hindering the implementation of smart agriculture practices.

Data management is, itself, a third challenge. Gathering and analyzing large amounts of data from sensors, drones, and other sources can be a challenge, and farmers may need to invest in specialized software and hardware to manage and analyze this data.

A fourth challenge is related to cybersecurity. The use of connected devices and data analytics in smart agriculture introduces new risks of cyberattacks and data breaches, which can compromise the security and privacy of farmers and their operations.

IoBees! USING TECHNOLOGY TO HELP POLLINATION

Pollination is evidently one of the most important activities in agriculture. Bees fly from one flower to another, pollinating them, and enabling fruits to raise. However, the number of bees in the world have reduced over the years, which can seriously threat the sustainability and food production chain. Technology has been used to overcome this challenge. A project in Europe named IoBee[1] creates and improves technological development to help bees to face their challenges.

As reported by Forbes, IoT technology can be used to helping to save bees and improve pollination and agriculture. One way this can be achieved is through smart beehives. By installing IoT sensors in beehives, environmental factors such as temperature, humidity, and sound can be monitored. This information can then be used to analyze bee behavior and hive health, and alert beekeepers if any issues arise. Tracking bee movements and behavior is another way in which IoT can help promote bee health and pollination.

By using IoT devices to track pollination patterns, valuable information can be gathered about the effectiveness of pollinators. This can help identify areas that may require additional pollinators, ultimately helping to improve crop yields and promote sustainable agriculture practices. Finally, IoT sensors can be used to monitor bee health and detect any signs of disease or stress. This enables beekeepers to take preventative measures to protect their hives and ensure the health of their bees, ultimately helping to save these vital creatures and support the health of our planet.

Source: https://www.forbes.com/sites/bernardmarr/2020/04/22/how-artificial-intelligence-iot-and-big-data-can-save-the-bees/?sh=3c207791d9ee

In addition, there is a challenge related to regulations. The use of advanced technologies and data analytics in agriculture may be subject to various regulations and laws, which can vary from one jurisdiction to another. This can create challenges for farmers and agribusinesses looking to implement smart agriculture practices.

Finally, there is a challenge related to the cost: implementing smart agriculture technologies can be expensive, particularly for small farmers and those in developing countries. The high cost of sensors, automation systems, and other technologies can be a barrier to adoption, especially for farmers who may already be facing economic challenges.

9.4 TYPES OF SENSORS THAT CAN BE USED IN SMART AGRICULTURE

There are many different sensors that can be used in smart agriculture to monitor and optimize crop growth, soil conditions, and environmental factors. Some common types of sensors used in smart agriculture include following:

1. **Temperature sensors:** These sensors can be used to measure the temperature of the air, soil, and water, which can help farmers to optimize irrigation and fertilization practices. Here are a few examples of temperature sensors with manufacturer names:
   a. Temperature data logger from MadgeTech
   b. TME temperature sensor from Microchip Technology
   c. LM35 temperature sensor from Texas Instruments
   d. TMP36 temperature sensor from Analog Devices
2. **Humidity sensors:** These sensors can measure the moisture content of the air and soil, which can help farmers to optimize irrigation and pest management practices. Here are a few examples of humidity sensors with manufacturer names:
   a. HIH-6130 humidity sensor from Honeywell
   b. SHT31 humidity sensor from Sensirion
   c. AM2320 humidity sensor from Aosong
   d. HDC2010 humidity sensor from Texas Instruments
3. **Light sensors:** These sensors can measure the intensity and spectrum of light, which can be used to optimize lighting conditions for greenhouse crops or to monitor the health of outdoor crops. Here are a few examples of light sensors with manufacturer names:
   a. TSL2561 light sensor from Adafruit Industries
   b. APDS-9960 light sensor from Avago Technologies
   c. BH1750 light sensor from ROHM Semiconductor
   d. TSL2591 light sensor from Adafruit Industries
4. **Soil moisture sensors:** These sensors can measure the moisture content of the soil, which can help farmers to optimize irrigation and fertilization practices.
5. **Soil nutrient sensors:** These sensors can measure the levels of various nutrients in the soil, such as nitrogen, phosphorus, and potassium, which can help farmers to optimize fertilization practices.
6. **Pest and disease sensors:** These sensors can detect the presence of pests and diseases in crops, which can help farmers to implement targeted pest management strategies.
7. **Weather sensors:** These sensors can measure various weather conditions, such as temperature, humidity, wind speed, and precipitation, which can help farmers to optimize irrigation and fertilization practices and to protect crops from extreme weather events.

9.5 SMART LIVESTOCK: CASE STUDY OF COMPOST BARN ON AGRIBUSINESS FROM BRAZIL

To ensure optimal animal welfare and agricultural productivity, it is crucial to monitor and control the environment in real time and make informed decisions that will positively impact animal health and production.

Farming and livestock production, key activities that require proper temperature, and humidity conditions to succeed, can be greatly impacted if these conditions are not met. For instance, if soil is too dry or wet, it can cause issues with pests and bacteria, while a hot or humid environment for animals can result in illness and decreased meat, milk, or egg production.

Smart livestock uses biosensors and software to monitor animal health and well-being during production, generating a large amount of data that require the use of intelligent systems to analyze and provide actionable insights to producers and managers.

This section examines a real case of a smart livestock system in milk production, using the compost barn principle.

Case study description (Gomes et al., 2023): The e-Livestock system for compost barn production is analyzed and deployed at the EMBRAPA – Gado de Leite experimental field. As part of a research project with Brazilian and international institutions aimed at improving the dairy cattle production system, the compost barn at EMBRAPA utilizes sensors to monitor various aspects of the animals, such as weight (kg), milk production (L), and mastitis (an inflammation of the mammary glands), as well as the ambient temperature (°C) and humidity (%) within the barn.

A compost barn is a structure designed for composting organic materials, such as animal manure, food waste, and plant matter. The open or partially open design of the barn provides proper ventilation and moisture control, and the organic matter inside is regularly turned and mixed to encourage breakdown by microorganisms. The composting process produces heat, which is used to warm the barn during winter, resulting in a nutrient-rich soil amendment that can be used for crop fertilization. The compost barn has several benefits, including reduced environmental impact, cost savings, and improved milk production quality.

Figure 9.2 illustrates the compost barn production system, which utilizes sensors in the environment and on the cows to monitor and adjust the animals' living

FIGURE 9.2   Example of compost barn e-sensors in livestock.

conditions. The cows are organized into lots based on their production, which can change over time, and housed in a shed with necessary infrastructure for their well-being. The compost barn is also well suited for cows that are susceptible to ticks, as the environment greatly minimizes its incidence and provides a pleasant environment for the animals.

The compost barn system uses technology to enhance animal welfare by monitoring and controlling relevant data through artificial intelligence (AI) mechanisms. The sensors collect data on the environment and the animals, which is processed and used to make automatic or human-triggered decisions to adjust the environment and improve the animals' welfare, such as turning a fan on or off or triggering the barn mixing. Environmental factors, such as heat, water, and humidity, can also be monitored and their impact on cow welfare observed to support decision-making. For example, early detection of mastitis can improve milk quality and prevent a drop in production. A system that can predict mastitis based on sensor data and the animal's history can aid in timely interventions and improve animal welfare and productivity. The following section details the structure of such a system.

9.5.1 THE TECHNOLOGICAL STRUCTURE OF THE SOLUTION

An architecture named as e-Livestock was developed to conform to the main principles of smart agribusiness solutions: monitor, management, and control. Figure 9.3 illustrates how e-Livestock architecture for dairy cattle was designed. The architecture is structured into tiers and modules, as follows.

- The *sensor tier* cleans, formats, and transforms the data that come from sensors attached to the animals. That tier collects IoT data generated by sensors deployed in the farm. Sensors collect information such as temperature and humidity. In addition, the sensor tier also handles different data

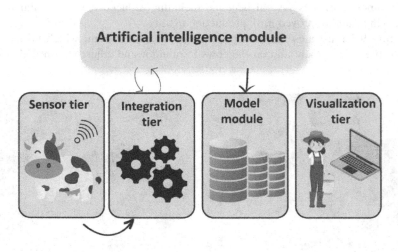

FIGURE 9.3   Dairy cattle architecture adapted.

formats generated by internal systems on farms. After cleaning, the data is formatted and sent to the integration tier to be integrated with external data sources and subsequently persisted in a database.

- The *integration tier* manages and processes the data collected, interacts with the AI module, and persists data in the model module. The integration tier is responsible for processing the collected data to be integrated with information from other sources and services, accessing context data, environmental information (temperature, humidity, or even weather forecast). The main advantage of this tier is that it can centralize and aggregate external information to the researcher to enrich the decision-making process.
- The *AI module* communicates with the integration tier, receiving the data already processed, then executing the most suitable intelligent algorithm for a given dataset. The AI module was developed to provide answers based on historical farm data. These answers can be milk production forecasts, mastitis type classification, and food consumption estimation. Based on historical data of the type of food, average consumption per animal, number of animals, and even time of year, it is possible to optimize the amount of adequate feed to be supplied to the animals. That action can avoid spending more than the necessary by buying commodities such as corn and soybeans, generating economic impacts on the farm. By preventing waste, we can contribute to a more sustainable farm. Algorithms can be added and even data from social media can also be used to get users' feedback on the quality of a particular farm-made product. Once the smart model is trained and ready to use, it requests input data from the API and sends the results generated by the model back to the API.
- The *model tier* deals with farm data and by integrating external data, it is possible to store the context of the data generated on the farm to extract information and enable the generation of dashboards. By capturing the data of an animal's feed and comparing it with its milk production, the system can identify the best diet to increase milk production, for example. The model tier also stores the metadata of the models, such as model accuracy, average errors, algorithm type, and input dataset used. In this way, it is possible to extract the provenance of the results, which can be used to make future decisions. For example, with the prediction of food consumption, the researcher can estimate the expected cost of purchasing inputs and plan storage according to the probability of consumption of animals indicated by the algorithm.
- The *visualization tier* allows the researcher to visualize the data in real time in a mobile or using a desktop, observing and analyzing information processed through a panel according to a time interval. The farmer can also analyze and interpret data at different granularities.

The use of this type of technology supports the farmers to answer questions such as:

i. Did the production of an animal that was eating correctly and had adequate weight decrease due to the occurrence of inflammation (mastitis)?

    ii. Was the animal's weight loss caused due to a change in diet that the animal did not adapt to?

   iii. Was an animal's weight loss due to a calving event?

   iv. Did the average production drop due to a temperature change?

    v. Did the temperature variation make the animal spend more energy maintaining body temperature instead of producing milk?

   vi. Did the average of mastitis cases grow, due to an increase in humidity, favoring the proliferation of environmental bacteria?, among other questions

As data is collected, processed, and made available on the dashboards, it becomes possible to constantly monitor the compost barn environment and detect events such as increased weight of animals, increase or decrease in milk production per batch, compost barn temperature, and humidity variation. By storing, training and testing intelligent models, information on average errors, model type, and input data, it was possible to analyze and discover the best predictive model for a given data set. Researchers could make more accurate planning and producers could investigate animals that were not producing as expected and improve the animal's life quality, food, and health. The producer could then prevent diseases and ensure animal welfare.

Using the monitoring capabilities, it was possible to identify a reduction in the milk production of one of the lots and with the provenance data, it was possible to investigate the reason for discarding the milk. During two months of monitoring, there was a high humidity level in the compost barn, which generated a proliferation of environmental bacteria. These bacteria caused mastitis in the animal, which consequently needed to be medicated and lost weight. Thus, the decision was made to discard the milk due to the medications. Another case was detecting the sudden increase in weight of one of the cows and its change of lot. A peak was observed in one of the lots by tracking the evolution of animal weights. With the provenance model, we identified that one of the cows in that batch had an insemination event, causing an increase in weight and, later, the decision to migrate a batch.

Furthermore, it was possible to estimate the milk production of this animal and monitor whether the projection would take place within the following months, using intelligent algorithms. The results of intelligent algorithms offered a future insight for production planning and animal management. For the context of this study, the neural network had a more accurate result, although, for other datasets, new tests need to be conducted to choose the best model. The results obtained cannot be generalized and additional studies will have to be conducted later.

As a result, the system provided decision support through data provenance with AI results were displayed in graphs and analyzed by the experimental field researchers. Once a decision was made or an event considered unusual was detected, the system could clarify the reason and the entire process that generated the decision.

9.5.2 AI in the Compost Barn System[2]

AI can support farmers in the decision-making process in animal management, aiming at their well-being. Integrating farm-generated data with AI make systems

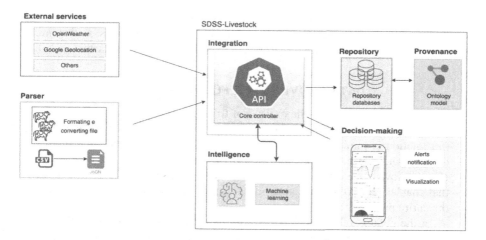

FIGURE 9.4 A livestock architecture overview to support Agriculture 4.0.

more complex as they drive future decisions. The literature shows the importance of interpreting data by intelligent systems to later support producers and managers in decision-making related to health and animal production. Using sensors and IoT devices on farms generates a large volume of data. Additionally, the soil assessment with drones and the use of sensors to capture real-time data at the field level are also important. These technologies, combined with cloud computing, can monitor agricultural components such as soils, plants, animals, weather, and other environmental conditions.

A possible solution is to support the collection, processing, and storage of data from a given rural environment and allow the use of intelligent algorithms. Figure 9.4 illustrates this proposed solution. In addition, it provides visualization mechanisms in dynamic dashboards in real time. This is an alternative organization for the system discussed in Section 9.5.1.

The architecture is divided into five main modules, as shown in Figure 9.4. The first module consists of cleaning, formatting, and transforming the data that arrive in the system. After cleaning, the data is formatted and sent to be integrated with external data sources and later persisted in a database. The integration module manages and processes the data collected by external services, interacts with the machine learning module, and persists the data in the repository module. The external services module comprises data sources external to the application. The repository module persists farm data, including metadata generated by the machine learning module so that you can track the prospective and retrospective provenance of prediction results. And finally, the visualization module supports decision-making. The modules and data flow at each step are described in more detail as follows.

- **Parser:** This module is responsible for cleaning, organizing, and formatting the data generated on the farm, whether data from sensors or those collected by rural producers. Sensor data are collected by body devices installed on animals or in the physical environment (context data). Parsers help by transforming the data that usually arrive in spreadsheet format (CSV or XLS) to JSON and is sent to the API, in the integration module, via HTTP request. The architecture was developed to deal with these data formats, as they are commonly exported by the sensors and internal systems of the farms.

DATA PROVENANCE FOR LIVESTOCK

Data provenance refers to the process of tracing and documenting the origin, ownership, and movement of data from its creation to its current state. This information is important for ensuring the authenticity, integrity, and quality of data, and can help to identify errors, inconsistencies, or potential fraud. Hence, provenance is intimately related to IoT, since sensors are the backbone that enables data to be collected, but their provenance needs to be assured in the entire production chain.

In the context of livestock, data provenance can be used to track the history of animals, including their birth date, breed, vaccinations, and medical treatment. This can be particularly useful for livestock farmers, veterinarians, and food producers who need to ensure the safety, quality, and traceability of their products. For example, by using data provenance, a farmer can trace the entire lifecycle of a cow, from its birth on the farm to its transport to the abattoir. Data can be stored in a blockchain (as discussed in Chapter 6) to that all the information is immutable and traceable. This information can be used to ensure that the cow was raised in a healthy environment, free from disease and harmful chemicals. It can also be used to track any medical treatment that the cow received, such as antibiotics or vaccinations, which can be important for food safety and regulatory compliance.

Data provenance can also help to ensure the integrity of animal welfare standards. By tracking the history of an animal, it is possible to ensure that it was raised in accordance with animal welfare regulations, such as free-range farming or humane slaughter practices.

- **Integration:** This module is responsible for receiving, processing, and integrating farm data. The main advantage of this module is being able to centralize and aggregate information external to the farm to enrich the decision-making process. The integration module can also aggregate information from external databases, services, and APIs, such as weather and weather data. It is in this module that communication with the machine learning module takes place. The machine learning module communicates directly with the integration module, receiving the already processed data. It then executes the most suitable intelligent algorithm for a given dataset (Neural Network or Random Forest, for example). It sends the results of predictions or classifications as a response to the API.
- **External services:** This module represents external services, databases, historical databases, social networks, and any external data sources that can be obtained to add value to the data collected by the parser module. According to the need, new sources can be easily coupled to the architecture through the integration module. By aggregating weather forecast data, for

example, it is possible to provide a new perspective for decision-making. This module is also related to provenance. Once the data is aggregated and stored, the recovery of metadata stored in the database enables the traceability of decisions by the provenance model. By recording information such as data source, sensor, and data type, it is possible to track and analyze the context of decisions that use this information. This module is responsible for receiving, processing, and integrating farm data. The main advantage of this module is being able to centralize and aggregate information external to the farm to enrich the decision-making process. The integration module can also aggregate information from external databases, services, and APIs, such as weather and weather data. It is in this module that communication with the machine learning module takes place. The machine learning module communicates directly with the integration module, receiving the already processed data. It then executes the most suitable intelligent algorithm for a given dataset (Neural Network or Random Forest, for example). It sends the results of predictions or classifications as a response to the API.

- **Repository:** This module represents the persistence of farm data and its provenance. Through the integration of external data, it is possible to store the context of the data generated on the farm to extract the retrospective provenance of this data and enable the generation of dashboards. The repository module also stores the metadata of the smart models, such as the model accuracy, the average of errors, the type of algorithm, and the input dataset used. In this way, it is possible to extract the provenance of the results of the smart models, which can be used to make future decisions. For example, with the prediction of food consumption, the farm can estimate the expected cost with the purchase of inputs according to the probability of consumption of the animals indicated by the algorithm.

- **Visualization:** The visualization module allows the producer to visualize the real time and historical data according to a time interval. In this way, it is possible to analyze and interpret the data at different granularities and visualize the expected production through the results of the machine learning module. In addition, a mobile application was developed as a visualization tool due to its ease of accessing dashboards and receiving alerts from the user.

9.6 POULTRY FARMING

According to the Internal Revenue Service of the Department of the Treasury of the United States, livestock includes "cattle, hogs, horses, mules, donkeys, sheep, goats, fur-bearing animals, and other mammals." The institution highlights (and it is also corroborated by the New World Encyclopedia) that the term "livestock" does not include poultry, chickens, turkeys, pigeons, geese, other birds, fish, shellfish, amphibians (frogs), and reptiles. Chickens are the birds that are most commonly associated with the term "poultry farming."

Poultry farming is substantial for several economies. For instance, in Brazil, chicken consumption in 2020 was around 45 kilograms of chicken meat per capita, increasing 5% compared to 2019 (42.84 kilograms). As these needs grow, IoT-based solutions (solutions that use technologies instead of human presence and influence) have been proposed to overcome the conventional ones (solutions that use human presence and influence instead of technologies) (Lopes, 2022). In comparison with technological and traditional solutions, the results show that the adoption of IoT-based solutions has increased productivity and optimized the use of farm resources (Phupattanasin and Tong, 2014).

Poultry production often faces difficulties due to the sensitivity that birds have to numerous factors of ambience in the aviary (Graciano Neto et al., 2022). The temperature is one of the factors that most impacts on losses of aviculture production which can be due, for instance, to the imbalance of the birds' thermal comfort zone. The imbalance of the environment is not only caused by the temperature. There are other variables that influence the welfare of poultry, such as lightning, harmful gases, air humidity, food quality, and clean water (Mumbelli et al., 2020).

As it was discussed, human mistakes in poultry monitoring activity can result in animal death. Thus, the systems that aim to monitor any livestock domain must be accurate (must monitor and display data closer to the reality) and fast, which allows us to classify this kind of system as critical (Lopes, 2022). IoT-based technologies support different aspects of livestock and agriculture context, such as monitoring essential data (temperature, humidity, and wind, among others) and transforming it into overall data, such as plots.

Solutions have been conceived to support such smart applications in poultry farming. A remarkable solution was recently conceived and deployed as a software-based system to monitor and support automated actions in aviary environments (Lopes et al., 2021). Figure 9.5 shows a conceptual illustration of it and a real picture from the deployed system. The system was developed in conformance to an architecture named as E-SECO (Ambrósio et al., 2021) involving sensors related to (i) luminosity, (ii) temperature, and (iii) air humidity linked to a LoRa module, which worked as

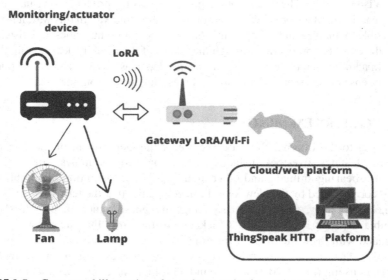

FIGURE 9.5   Conceptual illustration of a poultry monitoring and automated actions system.

the platform to which the sensors were connected, and the actuators, that is, devices linked to control a lamp (to increase the temperature and luminosity when required) and a fan (to refresh the environment, when needed). The LoRa platform with sensors and actuators was linked to a gateway so that all the data collected could be forwarded to a physically stored by a cloud/web-based service in a database (Gomes et al., 2021).

Once data was collected by sensors, they were transferred via the LoRa connection toward a gateway, from which data could be processed and decisions could be taken according to the thresholds established (such as an inferior threshold of 30 degrees Celsius), performing automated decisions on the environment, such as turning the lamp or fan on/off. The data collected by sensors had four purposes: to (i) serve as input for automated decision in situ, (ii) be stored in a remote database and to be delivered for further processing and analyses, (iii) be displayed to the farmers using the visualization, and (iv) be analyzed and mined to search for optimization patterns, such as to pre-defined triggers on temperature that exhibited better results over the system functioning eventually using the AI.

The system was deployed in a real farm in Posse, Goiás (Brazil) and worked for three days, from November 10, 2021 until November 13, 2021. Figure 9.6 shows the system and the chicken in the raising environment. A new experiment started on December 6, 2021 and ran for more time. During its operation, the system was well succeeded to monitor the aforementioned variables for 24 hours during 13 days. The lamp was turned-on most of the time to maintain the temperature above 30 degrees Celsius. The system operated the lamp (turning it on or off) 14 times, and the fan was not required to work since the maximum temperature set was 37 degrees Celsius and the local temperature was low due to recurrent rainfall in the farm region. Fifty birds were monitored during that period. There were two losses. The farmers classified the final result as effective, since they did not need to work directly on it during the period to monitor the considered variables. Those solutions can then be extended to other farms around the world.

FIGURE 9.6   The deployed system and the chicken in the raising environment.

9.7 FINAL REMARKS

This chapter discussed some challenges related to Agriculture 4.0 and how the IoT systems could support production systems in smart farms. Decisions made in agriculture need to be traceable due to the diversity of data and devices in different contexts. By using machine learning techniques, decision support systems can provide predictions. Also, combining and processing additional data sources and sensors can improve the accuracy of machine learning algorithms, leading to more accurate results. An example of using sensors in a compost barn was also presented, and an overview of an architecture that uses data provenance and machine techniques was discussed. This chapter discussed the adoption of technology for agriculture, livestock, and poultry farming. The next chapter addresses smart education in urban environments.

NOTES

1 https://io-bee.eu/
2 This is a more technical content related to software engineering. If the reader comes from another engineering, this section can be skipped.

REFERENCES

Ambrósio, L., Magaldi, H., David, J. M., Braga, R., Arbex, W., Campos, M., & Capilla, R. (2021). Enhancing the reuse of scientific experiments for agricultural software ecosystems. Journal of Grid Computing. https://doi.org/10.1007/s10723-021-09583-x

Bahlo, C., Dahlhaus, P., Thompson, H., & Trotter, M. (2019). The role of interoperable data standards in precision livestock farming in extensive livestock systems: A review. Computers and Electronics in Agriculture, 156, 459–466.

Bechhofer, S. (2009). OWL: Web ontology language. In Encyclopedia of Database Systems (pp. 2008–2009). Springer US, New York, NY.

Boichard, D., & Brochard, M. (2012). New phenotypes for new breeding goals in dairy 1655 cattle. The Animal Consortium, 6, 544–550.`

Buneman, P., Khanna, S., & Tan, C. (2001). "Why and where: A characterization of data provenance," In 8th Int. Conf. on Database Theory (pp. 4–6). London, UK.

Buneman, P., & Tan, W. (2019). Data provenance: What next? ACM SIGMOD Record, 47(3), 5–16.

Campos, M. M., Leao, J. M., Lima, J. A. M., & Machado, F. S. (2015). Precision technologies in food efficiency evaluation. Cadernos Técnicos de Veterinária e Zootecnia, 79, 73–85. (in Portuguese).

Cao, B., Plale, B., Subramanian, G., Robertson, E., & Simmhan. Y. (2009). Provenance information model of karma version 3. In 2009 Congress on Services-I (pp. 348–351). IEEE, Los Angeles, CA.

Cover, T., & Hart, P. (1967). "Nearest neighbor pattern classification," IEEE Transactions on Information Theory, 13(1), 21–27. https://doi.org/10.1109/TIT.1967.1053964

De Oliveira, D., Ogasawara, E., Baião, F., & Mattoso, M. (2010). SciCumulus: A lightweight cloud middleware to explore many task computing paradigm in scientific workflows. In 2010 IEEE 3rd International Conference on Cloud Computing (pp. 378–385). IEEE, Miami, FL.

Embrapa Gado de Leite. (2020). Brasil tem a primeira instalação de compost barn destinada a pesquisa. https://www.embrapa.br/busca-de-noticias/-/noticia/53360675/brasil-tem-a-primeira-instalacao-de-compost-barndestinada-apesquisa

Farooq, M. S., Riaz, S., Abid, A., Abid, K., & Naeem, M. A. (2019). A survey on the role of IoT in agriculture for the implementation of smart farming. IEEE Access, 7, 156237–156271.

Gil, Y., Deelman, E., Ellisman, M., Fahringer, T., Fox, G., Gannon, D., Goble, C., Livny, M., Moreau, L., & Myers, J. (2007). Examining the challenges of scientific workflows. Computer, 40(12), 24–32.

Gomes, J., David, J. M. N., Braga, R., Ströele, V., Arbex, W., Barbosa, B., Gomes, W., & Fonseca, L. (2021). "Architecture for decision support in precision livestock farming." In Proceedings of the 15th Brazilian e-Science Workshop (pp. 41–48). SBC.

Gomes, J., Lopes, V. C., Neto, V. V. G., De Oliveira, R. F., Kassab, M., David, J. M. N., ... & Arbex, W. (2022). Deriving experiments from E-SECO software ecosystem in the technology transfer process for the livestock domain. In Proceedings of the 10th IEEE/ACM International Workshop on Software Engineering for Systems-of-Systems and Software Ecosystems (pp. 1–8). Pittsburgh, PA.

Gomes, J., Esteves I., Graciano-Neto, V. V., David, J. M. N., Braga, R., Arbex, W., Kassab, M., & Oliveira, R. F. de Oliveira (2023). A scientific software ecosystem architecture for the livestock domain, Information and Software Technology, Volume 160, 107240, https://doi.org/10.1016/j.infsof.2023.107240.

Graciano Neto, V. V., Kassab, M., Lopes, V. C., Oliveira, R., & Bulcão-Neto, R. F. (2022, December). The state of IoT for agribusiness in Brazil. Computer, 55(12), 140–144.

Gruber, T. R. et al. (1993). A translation approach to portable ontology specifications. Knowledge Acquisition, 5(2), 199–220.

Gualdi, F., & Cordella, A. (2021). "Artificial intelligence and decision-making: The question of accountability." In Proceedings of the 54th Hawaii International Conference on System Sciences.

Karthick, G. S., Sridhar, M., & Pankajavalli, P. B. (2020). Internet of Things in animal healthcare (IoTAH): Review of recent advancements in architecture, sensing technologies and real-time monitoring. SN Computer Science, 1(5), 1–16.

Lakshmi, V., & Corbett, J. (2020). "How artificial intelligence improves agricultural productivity and sustainability: A global thematic analysis." In Proceedings of the 53rd Hawaii International Conference on System Sciences. Maui, HI.

Lopes, V. C. (2022). IoT-based architecture for monitoring and automated decision making in an aviary environment. Master's Thesis. Federal University of Goiás, Institute of Informatics (INF), Graduate Program in Computer Science, Goiânia, Brazil (p. 119).

Lopes, V., Oliveira, R., & Graciano Neto, V. (2021). Towards an IoT-based architecture for monitoring and automated decision-making in an aviary environment. In Anais do XIII Congresso Brasileiro de Agroinformática (pp. 320–328). SBC, Porto Alegre. https://doi.org/10.5753/sbiagro.2021.18404

Lopes, V. C., Silva, C., Gonçalves, D., Oliveira, R., Bulcão-Neto, R. F., Kassab, M., & Graciano-Neto, V. V. (2022). A systematic mapping study on IoT-based software systems for precision agriculture. International Journal of Computer Applications in Technology (IJCAT), 70(3–4), 155–170. Accepted to be published.

Moreau, L., & Missier, P. (2013). PROV-DM: The prov data model. W3C recommendation. World Wide Web Consortium (2013).

Mumbelli, A., Brito, R. C., Pegorini, V., & Priester, L. F. (2020, March). Low cost IoT-based system for monitoring and remote controlling aviaries. In 2020 3rd International Conference on Information and Computer Technologies (ICICT) (pp. 531–535). IEEE, San Jose, CA, USA.

Newlands, N. et al. (2019). "Deep learning for improved agricultural risk management." New World Encyclopedia. Livestock. https://www.newworldencyclopedia.org/entry/Livestock. Access on October 4th, 2022.

Paddock, Z. D. Energy expenditure in growing heifers with divergent residual feed intake phenotypes. Effects and interaction of metaphylactic treatment and temperament on receiving steers. PhD thesis. Texas A & M University.

Phupattanasin, P., & Tong, S. R. (2014). Applying information-centric networking in today's agriculture. Apcbee Procedia, 8, 184–188.

Pietersma, L., & Wade, K. M. (1998). A framework for the development of computerized management and control systems for use in dairy farming. Journal of Dairy Science, 81, 2962–2972.

Rumelhart, D., Hinton, G., & Williams, R. (1986). Learning representations by back-propagating errors. Nature, 323, 533–536. https://doi.org/10.1038/323533a0

Simmhan, Y. Plale, B., Gannon, D., & Marru, S. (2006). Performance Evaluation of the Karma Provenance Framework for Scientific Workflows. In International Provenance and Annotation Workshop (pp. 222–236). Springer.

Zhai, Z., Martínez, J. F., Beltran, V., & Martínez, N. L. (2020). Decision support systems for agriculture 4.0: Survey and challenges. Computers and Electronics in Agriculture, 170, 105256.

10 A Smarter Education for a Better Future

Opportunities and Challenges

10.1 INTRODUCTION

Smarter cities demand smarter citizens. Hence, smarter practices in education are demanded to potentialize citizens knowledge acquiring and usage. Particularly, the scenario of COVID-19 pandemic pressured educational institutions to accelerate a more intensive use of technology in their practices (Gonçalves et al., 2021). In a smart city, the education sector can be considered as one of the key components, and the integration of smart education practices can contribute to the overall smartness of the city. For example, a smart city can use technology to monitor the educational needs of its citizens, such as identifying areas with a high concentration of students and providing adequate resources for them. This, in turn, can lead to improved educational outcomes for the citizens, which can contribute to the overall development and progress of the city. Smart education can also benefit from the infrastructure and resources available in a smart city. For example, a smart city can provide high-speed internet connectivity, which can facilitate online learning and improve the accessibility of education. Smart cities can also provide state-of-the-art facilities and resources, such as smart classrooms and libraries, which can enhance the learning experience for students and provide them with the tools they need to succeed.

Smarter education refers to the intensive adoption of technology and innovative practices in the learning initiatives in all education levels, from primary school to universities. Smart education leverages technology to enhance the learning experience and improve educational outcomes, integrating various technologies such as artificial intelligence, machine learning, big data, Internet of Things (IoT) (as explored in Chapter 2), and augmented reality (AR) to create an immersive and interactive learning environment. This can make educational institutions to turn into smart schools and smart campuses. In that context, with the growing popularity of IoT usages across different sectors, educational institutions have explored the potential of incorporating IoT and other technologies in their educational activities to benefit students, instructors, and the education system as a whole.

IoT in particular has the potential to address a variety of modes, objectives, subjects, and perspectives in education. For instance, IoT devices can be used to monitor student attendance and in-class activities as proposed in the studies by Alotaibi (2015) and Jiang (2016). Similarly, IoT can be deployed in online education and online laboratories settings to monitor students and objects (Shi et al., 2010; Bin, 2012; Yin et al., 2012; Bisták, 2014; Lamri et al., 2014; Fernandez et al., 2015; Srivastava and Yammiyavar, 2016). IoT can also be used to turn the university environment smarter, such as by supporting the construction of meteorological stations connected to mobile applications that notify the users of how the climate in the university campus is. Then, the students can use appropriate clothes and protection means (such as an umbrella) to be protected against cold, rainy, or warm weather. Such instruments have low cost, potential of implementation in various regions, as well as it generates and stores the data to be used in climatology teaching and research (Abbate et al., 2022).

Moreover, IoT is being used in education for more practical scenarios as well. For example, Valpreda and Zonda (2016) proposed an IoT-based system to increase children's knowledge about agricultural food production and consumption. Similarly, Sula et al. (2013) and Sula et al. (2014) discussed using IoT to educate children with autism spectrum disorder.

In the United States (GSMA, 2012), Japan (Fuse et al., 2012), and the United Kingdom (UNESCO, 2014), educators are already incorporating IoT technologies in their pedagogical processes. Many academic institutions, such as the University of San Francisco (Nie, 2013), have incorporated IoT technology to enhance campus safety. Information technology and networking giant, Cisco, has also launched several IoT projects for the education sector through engagement with schools and students in a smart academic forum (Cisco, 2013).

This chapter presents an overview on the benefits and challenges faced in integrating IoT into the curriculum and educational environment. A summary of the already implemented tools and a list of challenges that need to be investigated are also presented. In addition, this chapter presents a tool for an adaptive learning experience that responds to a remote learner's emotions. This chapter answers questions related to the topic.

10.2 IoT IN EDUCATION: BENEFITS

10.2.1 PERCEPTION

Perception refers to a stakeholder's understanding and interaction with an IoT system in an educational setting. In this context, there are three primary stakeholders – instructors, students, and staff – each with a unique perception of the IoT technologies deployed in the field of education.

For instructors, IoT can aid in managing attendance, class sessions, and equipment availability for students. For example, RFID readers can be installed to identify students' electronic tags and track their activities (Jiang, 2016). Instructors can also use IoT to communicate with remote students and collect immediate feedback, as well as analyze sensor data to evaluate student performance, interest, and participation (Elyamany and AlKhairi, 2015; Ueda and Ikeda, 2016). Additionally, IoT can assist in confirming the identity of students and identifying students with special needs (Sula et al., 2014; Lenz et al., 2016).

From the student perspective, IoT provides opportunities for communication, collaboration, and remote access to learning resources. Students can communicate with classmates and share project data, as well as access adapted learning resources based on their knowledge level, interaction, and more (Peña-Ríos et al., 2012; Sula et al., 2013; Jeong et al., 2015).

For staff, IoT plays a role in tracking students, managing resources, and ensuring safety and security. For example, staff members can monitor and maintain student health, track academic resources, and manage events (Tan et al., 2014; Wang, 2014; Caţă, 2015). IoT can also assist in institutional energy management (Cisco, 2013).

Overall, the integration of IoT technologies in the field of education offers various benefits for each stakeholder group, improving the overall educational experience and facilitating efficient operations.

10.2.2 LEARNING PRINCIPLES

According to Ambrose et al. (2010), seven principles underlie effective learning, as illustrated in Figure 10.1: student prior knowledge, knowledge organization, motivation, mastery, practice and feedback, course climate, and self-directed learning. These principles are informed by research from various disciplines (CMU, 2017).

We can find that IoT technologies can positively impact each of these seven principles (Kassab et al., 2020).

Antle et al. (2016) propose the Story of Things (SoT) system, which allows children to learn about the story behind objects in their everyday life. The back-of-the-hand display is activated by stick-on finger sensors, allowing children to select from several stories stored in a crowdsourced database. This information is overlaid on the world through an AR contact lens to enhance "knowledge organization." Another

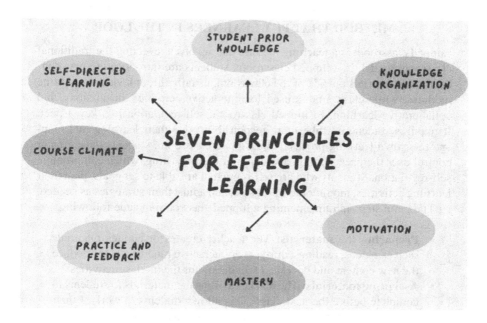

FIGURE 10.1   The seven principles for effective learning.

example of using IoT for "knowledge organization" is reported in the study by Gómez et al. (2013), which discuss a project that uses IoT to improve a child's attitude toward food by teaching about food consumption, production, and waste reduction.

In the work of Wan (2016), IoT is used to bridge communication between teachers and students and improve "motivation." Similarly, Elyamany and AlKhairi (2015) proposed an IoT-based system to analyze the impact of the physical classroom environment on students' focus, with the goal of improving the "course climate." Other studies with similar objectives include the works of Uzelac et al. (2015); Richert et al. (2016); Ueda and Ikeda (2016).

For example, Sula et al. (2014) propose a smart assistive environment system that uses a Heuristic Diagnostic Teaching (HDT) process to identify each student's learning abilities and creativity in mathematics. The system provides personalized "practice and feedback" by utilizing a computer, sensors, RFID tag reader, and SmartBox device to support learning for students with autism spectrum disorder. Other studies (Wang, 2010; Chen et al., 2011; Peña-Ríos et al., 2012; Sula et al., 2013; Ma and Li, 2013; Möller et al., 2013; Meda et al., 2014; Sula et al., 2014; Jeong et al., 2015; Murphy et al., 2015; Lenz et al., 2016; Gómez et al., 2016) discuss using IoT to provide a personalized learning experience. "Course climate" in foreigner language courses can be improved by using tools that enable to use music as a resource to learn (Graciano Neto et al., 2019), and smart applications can help students in their self-regulation during "self-directed learning," which can also be leveraged in universities during the adoption of flipped classroom approaches (Gonçalves et al., 2020).

FLIPPED CLASSROOM: THE EDUCATIONAL METHOD THAT PUT LEARNERS IN THE LOOP

Flipped classroom is a teaching method that involves reversing the traditional classroom model. In a flipped classroom, students are introduced to new concepts and materials outside of the classroom, usually through videos or online modules, while class time is used for interactive activities, discussions, and collaborative learning. In flipped classroom, self-regulation is a key aspect. It requires students to take more responsibility for their learning and manage their own learning process. In a flipped classroom, students have greater control over the pace, timing, and depth of their learning, which can promote self-regulation. Students who are self-regulated are able to set goals, plan their learning activities, monitor their progress, and adjust their strategies as needed.

The main steps in implementing a flipped classroom include following:

- **Preparing the materials:** The teacher creates or curates materials such as videos, readings, quizzes, or interactive modules that introduce the new content and provide a foundation for the in-class activities.
- **Assigning materials:** The teacher assigns the materials to students to complete before the next class. This allows students to work at their own pace and review the material as many times as needed.

- **Classroom activities:** During class time, the teacher facilitates activities such as group discussions, problem-solving, projects, or hands-on experiments that reinforce and apply the concepts learned in the pre-class materials. The teacher provides individualized guidance and feedback to help students achieve their learning goals.
- **Assessment:** The teacher assesses student learning through formative assessments such as quizzes, polls, or discussions during class time and summative assessments such as examinations or projects at the end of the unit.
- **Reflection:** The teacher and students reflect on the learning process and the effectiveness of the flipped classroom approach. They identify areas for improvement and adjust their teaching and learning strategies accordingly.

The flipped classroom approach can be used in any subject area and at any level of education. It is designed to enhance student engagement, deepen understanding, and promote active learning.

10.2.3 DELIVERY MODE

Education delivery can occur through face-to-face, remote, or hybrid modes. According to the systematic literature review (SLR) conducted in the study by Kassab et al. (2020), the selected studies were evenly distributed between face-to-face and online settings, with IoT being discussed as a technology that can be used in all three modes. Examples of how IoT can be used in face-to-face education include utilizing sensors and RFID technology to personalize feedback for students with learning disabilities in math, as proposed in the study by Sula et al. (2014), and analyzing the impact of physical environmental parameters on students' focus in the classroom through an IoT-based system, as discussed in the study by Elyamany and AlKhairi (2015). Examples of how IoT can be used in online education include improving communication between teachers and students using IoT, as proposed in the study by Wan (2016) and enabling children to learn about the stories behind objects through an IoT-based SoT system, as discussed in the study by Antle et al. (2016).

In hybrid education, IoT can be used to create a seamless integration of face-to-face and remote learning experiences, combining the best of both worlds. For example, utilizing IoT-enabled devices to enhance collaboration and interaction between students and teachers in real time, whether they are in the same room or at a distance.

10.3 IoT IN EDUCATION: CHALLENGES

In the area of IoT in education, there are several challenges that need to be addressed in order to achieve effective and efficient use of the technology. In the work of Kassab et al. (2020), the authors discussed three significant challenges: security, scalability, and humanization. In terms of security, the protection of sensitive educational data and personal information is a top priority. Scalability is another challenge, as the technology must be able to accommodate the growing needs of an educational system, while still providing an efficient and effective user experience. On the other hand, humanization refers to ensuring that the technology is user-friendly and intuitive, which can be a challenge in the implementation of IoT in the education sector. These three challenges are further discussed in this section.

10.3.1 SECURITY

Security is a critical aspect of education and has become even more crucial with the increased use of IoT technology (Georgescu and Popescu 2015). The use of IoT in education has increased the vulnerability of educational systems to cyber-attacks, and the number of attacks is expected to rise in the future (Weber 2010). The interconnectivity of a large number of devices in the educational system makes it difficult to simultaneously protect all devices over air transmission, leading to cascade failures (Georgescu and Popescu, 2015). The higher education sector is particularly prone to data breaches, with 727 reported breaches in academic institutions in the US between 2005 and 2014, second only to the healthcare sector (Grama, 2014). Nearly 7% of all academic institutions in the US have experienced at least one data breach, with one-third of these institutions reporting more than one breach and 6% reporting five or more breaches. Hacking and malware, where an outside party accessed records through direct entry, malware, or spyware, accounted for the largest proportion of reported breaches at 36% (Grama, 2014).

The collection of various data by devices used in a specialized IoT raises questions about who owns the data and where it goes, requiring answers from the legal profession, government entities overseeing education, and educational standards groups (Laplante et al., 2015). To better understand information security issues in higher education, the Higher Education Information Security Council (HEISC) was established in July 2000 to provide coordination and support for information security governance, compliance, and data protection and privacy to higher education institutions (Grama and Vogel, 2016). The HEISC has identified "planning for and implementing next-generation security technologies" as one of its three strategic information security with a focus on the increasing concerns of IoT (Grama and Vogel, 2016).

10.3.2 SCALABILITY

According to Zhang (2016), by incorporating sensors into both front field environments and terminal devices, an IoT network can gather a wealth of sensor data that reflects real-time environmental conditions and events/activities in the front field. Advanced data mining techniques can then be applied to extract valuable business insights from these data. The high volume of data collected on an elementary event level in a 24/7 mode and its distinct access pattern compared to traditional business data has prompted the development of new data management solutions, such as NoSQL databases and MapReduce distributed computing frameworks.

In the education domain, IoT integration also leads to the generation of large amounts of data, necessitating the analysis and treatment of these data to capture information and trends. This scalability issue has been addressed in many studies, such as Jagtap et al. (2016), who propose a design for a "social recommender" system based on Hadoop and its parallel computing platform to provide personalized content to students based on analyzing a large volume of student data and activity. Mehmood et al. (2017) discuss the scalability concern when designing a learning management system.

However, the cost of IoT technology in education is also a significant concern, as it raises the question of whether it will increase the division between those who can afford these technologies and those who cannot. If affordable education for all is a priority, how will schools be able to purchase and maintain these devices? The financial aspect of IoT in education is discussed in some studies; such as the works of Maleko et al. (2012), Putjorn et al. (2015), Georgescu and Popescu (2015), and Pruet et al. (2015).

10.3.3 HUMANIZATION

The moral role that IoT may play in human lives, particularly in terms of personal control, raises questions. Applications in IoT involve interactions between not just computers but also humans. The success of IoT ultimately depends on the humanization of connected technologies (Tech, 2015). This shift toward increased dependence on technology may result in reduced autonomy for individuals and greater power for corporations driven by financial gain (Gubbi et al., 2013). The dehumanization of human interaction with machines is a concern that has been discussed in two studies (Murphy et al., 2015; Lenz et al., 2016). Then, a process of "rehumanization" is taking place in a movement called as Education 5.0 (Seeling et al., 2022).

Face-to-face interaction between students has been shown to benefit their social skills and contribute to character building. The increased use of IoT in the education system raises the possibility of a partial loss of the social aspect of going to school. However, IoT can also be beneficial for special needs students in virtual learning environments, providing support through features such as the ability to repeat experiments and access to prejudice-free performance evaluations through anonymization (Lenz et al., 2016). For students with dyslexia and dyscalculia, for example, anonymization and features such as auto-correct may improve biases in their scores and lead to a more fair assessment by teachers.

EDUCATION 5.0

The concepts of Industry 4.0, a movement about intensive use of IoT in industry, has also been migrated to education. The emergence of Robotics, Augmented and Virtual Reality, Artificial Intelligence, Big Data, and Automation has transformed education around the world. However, a new movement has tried to reascend the position of humans in the process, "rehumanizing" it. It has been named as Education 5.0.

Education 5.0 is an emerging educational paradigm that builds on the principles of previous educational paradigms and incorporates new technologies and socioeconomic realities. Education 5.0 is characterized by its focus on personalization, lifelong learning, and global citizenship, being a holistic approach to education that aims to develop the whole person, including their cognitive, emotional, social, and spiritual aspects. Education 5.0 emphasizes the development of skills such as critical thinking, creativity, collaboration, communication, and digital literacy. It represents a paradigm shift in education toward a more personalized, flexible, and inclusive approach that aims to prepare individuals for the challenges of the 21st century.

10.4 MONITORING EMOTIONAL STATE OF ONLINE LEARNER: A TOOL

The incorporation of IoT and Big Data analytics into education presents new opportunities to address the presence of emotions in learning. This combination has the potential to improve motivation and satisfaction among online learners by detecting and reacting to their emotional state in real time. The use of devices such as webcams and fitness monitors can provide affordable and non-intrusive means of capturing data such as facial expressions, eye tracking, and heart rate.

In response to this, a learning management system (LMS) was developed by the research team led by the second author of this book using Python, JQuery, material for UI, and WebSocket for communication with a web server. The system utilizes the Tornado framework for web service, OpenCV, Keras CNN model, and TensorFlow API for real-time face detection and emotion/gender classification. The system captures the learner's expressions and categorizes them into seven emotional states (angry, disgust, fear, happy, sad, surprise, and neutral) based on gender and the content being viewed. The process of capturing and processing emotional states is often called as affective computing (Reis et al., 2015).

Once the learner logs into the website, a pop-up message alerts them that their webcam will be turned on. As they navigate through the online materials, the system captures their facial expressions every 100 ms (Figure 10.2). This approach can provide valuable insights into the well-being of the students and contribute to the improvement of the feedback provided.

The system is designed to periodically evaluate the learner's expressions, with the evaluation frequency set to 10 s and customizable by a course administrator. The system compares the learner's expressions with the customized parameters while they navigate the course content and identifies a match if a certain percentage of emotional expressions (such as sad or happy) are detected within the 10-s timeframe. In case of a match, the learner is prompted with the option to access supplemental learning materials tailored to their current state and the course content they are interacting with (as illustrated in Figure 10.3).

The system also includes an "aggregator" component that sends periodic updates on the learner's profile and the overall state of the online class to the instructor. The

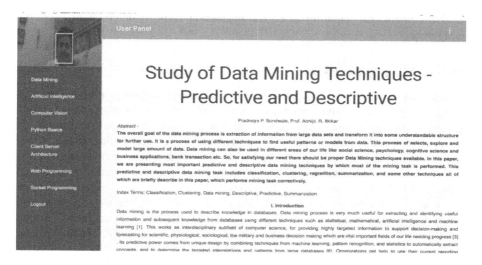

FIGURE 10.2 A screenshot from the tool at runtime while a remote learner navigates through online content.

captured data can benefit both the individual learner and the entire class. Although the first phase of the project successfully integrated a facial expression API, the development team is currently working on integrating eye-gaze tracking and heart-beats monitoring APIs. These additional forms of data collection will allow for a more precise determination of the learner's state while they are engaging with the

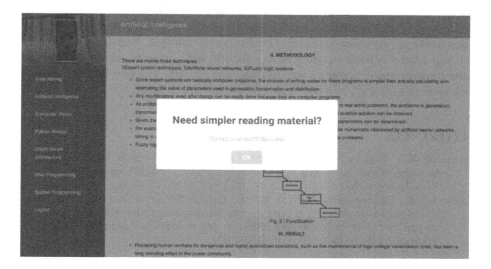

FIGURE 10.3 A screenshot from the tool at runtime. The tool detects a particular percentage of emotional expressions and prompts the learner with supplemental learning materials corresponding to the content.

course content, participating in online discussion forums, or completing course assignments.

10.5 FINAL REMARKS

There is an active effort toward implementing scenarios for integrating IoT in education; such as smart environment, e-learning and education management, virtual learning, remote laboratories, human motion capture, ubiquitous learning, AR e-learning, wireless robotic educational platform, cybersecurity laboratory, attendance system based on IoT, and psychological health education service based on IoT.

However, the literature review also identified significant research gaps and challenges related to IoT education and its derivatives. The gaps are driven by the large number of technologies, diverse application domains, applicable standards, and cross-disciplinary nature of IoT education. For example, there are questions on what are the right devices and processing components for IoT pedagogy? There is a need for research to understand a sufficient and cost-effective subset of technologies for the development and delivery of IoT curricula and experimental laboratories.

There are also concerns on how the instructors, staff, and students are going to connect and use the IoT network for teaching and learning. Smartphones/mobile devices are currently found to be the most commonly used devices, followed by sensors/RFID technologies (Kassab et al., 2020). Further research is needed toward defining a safe and efficient networking structure for IoT program delivery, as there are significant security, safety, and privacy issues surrounding IoT systems.

It is worth to mention that IoT systems are highly complex, and many applicable standards apply to them. Any IoT-based curricula must address the identification of appropriate standards and the harmonization of differences. The advances in technology are set to transform the education domain, but adding these standards to the education system could also increase its complexity. The US National Institute of Standards and Technologies Special Publication SP 800-183, "Networks of Things," should form the basic building blocks for IoT and a framework for researching and developing IoT scenarios for the education domain. The next and last chapter addresses a case study of a real deploy of a smart city in Brazil.

REFERENCES

Ambrose, S. A., Bridges, M. W., DiPietro, M., Lovett, M. C., & Norman, M. K. (2010). How Learning Works: Seven Research-Based Principles for Smart Teaching. John Wiley & Sons, Hoboken, NJ.

Abbate, J. K., Graciano Neto, V. V., & Barros, J. R. (2022). Experimento de baixo custo: Contribuição para coleta de dados e monitoramento para estudos climáticos (Low-cost experiment: Contribution to data collection and monitoring for climate studies – In Portuguese). Ateliê Geográfico, 16(2), 104–121.

Alotaibi, S. J. (2015). Attendance system based on the Internet of Things for supporting blended learning. In 2015 World Congress on Internet Security (WorldCIS) (pp. 78–78). Dublin, Ireland.

Alvarez, I. B., Silva, N. S. A., & Correia, L. S. (2016). Cyber education: Towards a pedagogical and heuristic learning. ACM SIGCAS Computers and Society, 45(3), 185–192.

Antle, A. N., Matkin, B., & Warren, J. (2016). The Story of Things: Awareness through Happenstance Interaction. In International Conference on Interaction Design and Children (pp. 745–750). Manchester, United Kingdom.

Bin, H. (2012). The design and implementation of laboratory equipments management system in university based on internet of things. In International Conference on Industrial Control and Electronics Engineering (pp. 1565–1567). Xi'an, China.

Bisták, P. (2014). Mobile web application for group of remote laboratories using middleware. In IEEE 12th IEEE International Conference on Emerging eLearning Technologies and Applications (ICETA) (pp. 51–56). Slovakia.

Borges, V., Sawant, R., Zarapkar, A., & Azgaonkar, S. (2011). Wireless automated monitoring system for an educational institute using Learning Management System (MOODLE). In International Conference of Soft Computing and Pattern Recognition (SoCPaR) (pp. 231–236). Dalian, China.

Brady, C., Weintrop, D., Gracey, K., Anton, G., & Wilensky, U. (2015). The CCL-parallax programmable badge: Learning with low-cost, communicative wearable computers. In Conference on Information Technology Education (pp. 139–144). Chicago, IL, USA.

Caţă, M. (2015). Smart university, a new concept in the Internet of Things. In Roedunet International Conference-Networking in Education and Research (roedunet ner) (pp. 195–197). Craiova, Romania.

Chen, J., Man, H., Jin, Q., & Huang, R. (2011). Goal-driven navigation for learning activities based on process optimization. In International Conference on Internet of Things and 4th International Conference on Cyber, Physical and Social Computing (pp. 389–395). Espoo, Finland.

Chunxia, J. (2015). Laboratory management of the internet based on the technology of Internet of Things. In AASRI International Conference on Industrial Electronics and Applications (IEA 2015). London, UK.

Cisco. (2013). Education and the Internet of Everything – How Ubiquitous Connectedness Can Help Transform Pedagogy. https://www.coursehero.com/file/19548989/education-internet/

CMU. (2017). Principles of Learning. https://www.cmu.edu/teaching/principles/learning.html

De la Guía, E., Camacho, V. L., Orozco-Barbosa, L., Luján, V. M. B., Penichet, V. M., & Pérez, M. L. (2016). Introducing IoT and wearable technologies into task-based language learning for young children. IEEE Transactions on Learning Technologies, 9(4), 366–378.

De la Torre, L., Sánchez, J. P., & Dormido, S. (2016). What remote labs can do for you. Physics Today, 69(4), 48–53.

DeFranco, J., Kassab, M., & Voas, J. (2018). How do you create an Internet of Things workforce? IT Professional, 20(4), 8–12.

Dybå, T., & Dingsøyr, T. (2008). Empirical studies of agile software development: A systematic review. Information and Software Technology, 50(9–10), 833–859.

Elyamany, H. F., & AlKhairi, A. H. (2015). IoT-academia architecture: A profound approach. In IEEE/ACIS 16th International Conference on Software Engineering, Artificial Intelligence, Networking and Parallel/Distributed Computing (SNPD) (pp. 1–5). Takamatsu, Japan.

Fernandez, G. C., Ruiz, E. S., Gil, M. C., & Perez, F. M. (2015). From RGB led laboratory to servomotor control with websockets and IoT as educational tool. In Proceedings of 2015 12th International Conference on Remote Engineering and Virtual Instrumentation (REV) (pp. 32–36). Bangkok, Thailand.

Fuse, M., Ozawa, S., & Miura, S. (2012). Role of the Internet for risk management at school. In International Conference on Information Technology Based Higher Education and Training (ITHET) (pp. 1–6). Istanbul, Turkey.

García, M. L., Fernandez, G. C., Ruiz, E. S., Martín, A. P., & Gil, M. C. (2013). Rethinking remote laboratories: Widgets and smart devices. In IEEE Frontiers in Education Conference (FIE) (pp. 782–788). Oklahoma City, Oklahoma, USA.

Gartner. (2017). The Internet of Things (IoT) Is a Key Enabling Technology for Digital Businesses. http://www.gartner.com/technology/research/internet-of-things/

Georgescu, M., & Popescu, D. (2015). How could Internet of Things change the E-learning environment. In The 11th International Scientific Conference eLearning and Software for Education. Bucharest, Romenia.

Gómez, J. E., Huete, J. F., & Hernandez, V. L. (2016). A contextualized system for supporting active learning. IEEE Transactions on Learning Technologies, 9(2), 196–202.

Gómez, J., Huete, J. F., Hoyos, O., Perez, L., & Grigori, D. (2013). Interaction system based on internet of things as support for education. Procedia Computer Science, 21, 132–139.

Gonçalves, A. C., Ferreira, D. J., & Graciano Neto, V. V. (2021). Computer programming Teachers' challenges for promoting students regulation on flipped learning activities during COVID-19. Revista Diálogo Educacional, 21(71), 1820–1838.

Gonçalves, A. C., Neto, V. V. G., Ferreira, D. J., & Silva, U. F. (2020). Flipped classroom applied to software architecture teaching. In 2020 IEEE Frontiers in Education Conference (FIE) (pp. 1–8). IEEE.

Graciano Neto, V. V., Siqueira, S., & Borim, M. C. (2019). Find me a song and add the blanks: Supporting teachers to retrieve lyrics to English listening lessons. In 2019 IEEE 19th International Conference on Advanced Learning Technologies (ICALT) (Vol. 2161, pp. 267–271). IEEE. Maceió, Brazil.

Grama, J. (2014). Just in Time Research: Data Breaches in Higher Education. EDUCAUSE.

Grama, J., & Vogel, V. (2016). The 2016 Top 3 Strategic Information Security Issues. https://er.educause.edu/articles/2016/1/the-2016-top-3-strategic-information-security-issues

GSMA. (2012). Mobile Education in the United States. https://www.gsma.com/iot/mobile-education-in-the-united-states/

Gubbi, J., Buyya, R., Marusic, S., & Palaniswami, M. (2013). Internet of Things (IoT): A vision, architectural elements, and future directions. Future Generation Computer Systems, 29(7), 1645–1660.

Gul, S., Asif, M., Ahmad, S., Yasir, M., Majid, M., Malik, M., & Arshad, S. (2017). A survey on role of internet of things in education. International Journal of Computer Science and Network Security, 17(5), 159–165.

Guyatt, G., Rennie, D., Meade, M. O., & Cook, D. J. (2003). Users' guides to the medical literature. Essentials of evidence-based clinical practice. Health Information and Libraries Journal, 20(2; SUPP/1), 79–79.

Ha, I., & Kim, C. (2014). The research trends and the effectiveness of smart learning. International Journal of Distributed Sensor Networks, 10(5), 537346.

Haiyan, H., & Chang, S. (2012). The design and implementation of ISIC-CDIO learning evaluation system based on Internet of Things. In World Automation Congress (pp. 1–4).

Han, W. (2011). Research of intelligent campus system based on IOT. In Advances in Multimedia, Software Engineering and Computing, vol. 1 (pp. 165–169). Springer.

He, B.-X., & Zhuang, K.-J. (2016). Research on the intelligent information system for the multimedia teaching equipment management. In International Conference on Information System and Artificial Intelligence (ISAI) (pp. 129–132). Hong Kong, China.

Heng, Z., Yi, C. D., & Zhong, L. J. (2011). Study of classroom teaching aids system based on wearable computing and centralized sensor network technique. In International Conference on Internet of Things and 4th International Conference on Cyber, Physical and Social Computing (pp. 624–628). Washington DC, United States.

Hernandes, E., Zamboni, A., Fabbri, S., & Thommazo, A. D. (2012). Using GQM and TAM to evaluate StArt-a tool that supports systematic review. CLEI Electronic Journal, 15(1), 3–3.

Hui, Y., & Haiyan, H. (2011). Research and realization on CDIO teaching experimental system based on RFID technique of Internet of Things. In International Conference on Mechatronic Science, Electric Engineering and Computer (MEC) (pp. 841–844).

Inayat, I., Salim, S. S., Marczak, S., Daneva, M., & Shamshirband, S. (2015). A systematic literature review on agile requirements engineering practices and challenges. Computers in Human Behavior, 51, 915–929.

International Data Corporation, "Steady Commercial and Consumer Adoption will Drive Worldwide Spending on the Internet of Things to $1.1 Trillion in 2023, According to a New ICD Guide", June 13, 2019, https://www.businesswire.com/news/home/20190613005028/en/Steady-Commercial-and-Consumer-Adoption-Will-Drive-Worldwide-Spending-on-the-Internet-of-Things-to-1.1-Trillion-in-2023-According-to-a-New-IDC-Spending-Guide retrieved 10/18/19.

Jagtap, A., Bodkhe, B., Gaikwad, B., & Kalyana, S. (2016). Homogenizing social networking with smart education by means of machine learning and Hadoop: A case study. In International Conference on Internet of Things and Applications (IOTA) (pp. 85–90). Pune, India.

Jeong, K., Kim, H.-S., & Chong, I. (2015). Knowledge driven composition model for WoO based self-directed smart learning environment. In International Conference on Information Networking (ICOIN) (pp. 537–540). Siem Reap, Cambodia.

Jiang, Z. (2016). Analysis of student activities trajectory and design of attendance management based on Internet of Things. In 2016 International Conference on Audio, Language and Image Processing (ICALIP) (pp. 600–603). Shanghai, China.

Jurkovičová, L., Červenka, P., Hrivíková, T., & Hlavatý, I. (2015). E-learning in augmented reality ` utilizing iBeacon technology. In Proceedings of the European Conference on E-Learning (pp. 170–178). Hatfield, United Kingdom.

Kane, P., Duda, M., Farrell, S., Jeffords, J., Kimsey, T., Rucinski, A., Zhong, J. (2013). Disruptive engineering training and education based on the Internet of Things. In IEEE International Conference on Teaching, Assessment and Learning for Engineering (TALE) (pp. 632–636).

Kassab, M., DeFranco, J., & Laplante, P. (2020). A systematic literature review on Internet of Things in education: Benefits and challenges. Journal of Computer Assisted Learning, 36(2), 115–127.

Keele, S., et al. (2007). Guidelines for performing systematic literature reviews in software engineering (Tech. Rep.). Technical report, Ver. 2.3 EBSE Technical Report. EBSE. Kitchenham, B., & Charters, S. 2007. Guidelines for performing systematic literature reviews in software engineering.

Lamri, M., Akrouf, S., Boubetra, A., Merabet, A., Selmani, L., & Boubetra, D. (2014). From local teaching to distant teaching through IoT interoperability. In 2014 International Conference on Interactive Mobile Communication Technologies and Learning (IMCL2014) (pp. 107–110). Thessaloniki, Greece.

LAPES. (2014). StArt (state of the art through systematic review tool). https://edisciplinas.usp.br/mod/url/view.php?id=1014403

Laplante, P. A. (2017). Requirements Engineering for Software and Systems. Auerbach Publications. Boca Raton, FL.

Laplante, P., Laplante, N., & Voas, J. (2015). Considerations for healthcare applications in the Internet of Things. Rel. Dig., 61(4), 8–9.

Lei, L., Dai, Q., Wang, M., Liu, Q., & Xiao, M. (2016). The research and implementation of engineering training system based on mobile Internet of Things. In IEEE International Conference on Consumer Electronics-China (ICCE-China) (pp. 1–4). Guangzhou, China.

Lenz, L., Pomp, A., Meisen, T., & Jeschke, S. (2016). How will the Internet of Things and Big Data analytics impact the education of learning-disabled students? A concept paper. In 3rd MEC International Conference on Big Data and Smart City (ICBDSC) (pp. 1–7). Muscat, Oman.

Li, J., & Li, X. (2014). Design of management system for teaching equipment based on the Internet of Things. Contemporary Logistics (15), 33.

Ma, G., & Li, Y. (2013). Application of IOT in information teaching of ethnic colleges. In Proceedings of the 2013 International Conference on Information, Business and Education Technology (ICIBET-2013). (pp. 403–406). Atlantis Press, Beijing, China.

Maleko, M., Hamilton, M., & D'Souza, D. (2012). Novices' perceptions and experiences of a mobile social learning environment for learning of programming. In Proceedings of the 17th ACM Annual Conference on Innovation and technology in Computer Science Education (pp. 285–290). Bologna Italy.

Meda, P., Kumar, M., & Parupalli, R. (2014). Mobile augmented reality application for Telugu language learning. In IEEE International Conference on MOOC, Innovation and Technology in Education (MITE) (pp. 183–186). Amritsar, India.

Mehmood, R., Alam, F., Albogami, N. N., Katib, I., Albeshri, A., & Altowaijri, S. M. (2017). UTiLearn: A personalised ubiquitous teaching and learning system for smart societies. IEEE Access, 5, 2615–2635.

Miglino, O., Di Fuccio, R., Di Ferdinando, A., & Ricci, C. (2014). BlockMagic, A hybrid educational environment based on RFID technology and Internet of Things concepts. In International Internet of Things Summit (pp. 64–69). Rome, Italy.

Möller, D. P., Haas, R., & Vakilzadian, H. (2013). Ubiquitous learning: Teaching modeling and simulation with technology. In Grand Challenges on Modeling and Simulation Conference (p. 24). Toronto, Ontario, Canada.

Murphy, F. E., Donovan, M., Cunningham, J., Jezequel, T., García, E., Jaeger, A., & Popovici, E. M. (2015). i4Toys: Video technology in toys for improved access to play, entertainment, and education. In IEEE International Symposium on Technology and Society (ISTAS) (pp. 1–6). Dublin, Ireland.

Nie, X. (2013). Constructing smart campus based on the cloud computing platform and the Internet of Things. In 2nd International Conference on Computer Science and Electronics Engineering. Khartoum, Sudan.

Pei, X. L., Wang, X., Wang, Y. F., & Li, M. K. (2013). Internet of Things based education: Definition, benefits, and challenges. Applied Mechanics and Materials, 411, 2947–2951.

Peña-Ríos, A., Callaghan, V., Gardner, M., & Alhaddad, M. J. (2012a). Towards the next generation of learning environments: An InterReality learning portal and model. In Eighth International Conference on Intelligent Environments (pp. 267–274). Guanajuato, Mexico.

Peña-Ríos, A., Callaghan, V., Gardner, M., & Alhaddad, M. J. (2012b). Remote mixed reality collaborative laboratory activities: Learning activities within the InterReality Portal. In IEEE/WIC/ACM International Joint Conferences on Web Intelligence and Intelligent Agent Technology-Volume 03 (pp. 362–366). Singapore.

Petersen, K., Feldt, R., Mujtaba, S., & Mattsson, M. (2008). Systematic mapping studies in software engineering. In EASE (Vol. 8, pp. 68–77).

Petrov, C., "47 Stunning Internet of Things Statistics 2020 [The Rise of IoT]," tech jury, 10/13/20, https://techjury.net/blog/internet-of-things-statistics/#gref, retrieved 10/18/20.

Pruet, P., Ang, C. S., Farzin, D., & Chaiwut, N. (2015). Exploring the Internet of "Educational Things" (IoET) in rural underprivileged areas. In 12th International Conference on Electrical Engineering/Electronics, Computer, Telecommunications and Information Technology (ECTI-CON) (pp. 1–5). Hua Hin, Thailand.

Putjorn, P., Ang, C. S., & Farzin, D. (2015). Learning IoT without the I-educational Internet of Things in a developing context. In Workshop on Do-It-Yourself Networking: An Interdisciplinary Approach (pp. 11–13).

Qi, A.-q., & Shen, Y.-j. (2011). The application of Internet of Things in teaching management system. In International Conference of Information Technology, Computer Engineering and Management Sciences (Vol. 2, pp. 239–241).

Reis, R. C. D., Rodriguez, C. L., Lyra, K. T., Jaques, P. A., Bittencourt, I. I., & Isotani, S. (2015). Affective states in CSCL environments: A Systematic mapping of the literature. In 2015 IEEE 15th International Conference on Advanced Learning Technologies (pp. 335–339). IEEE. Hualien, Taiwan.

Richert, A., Shehadeh, M., Plumanns, L., Groß, K., Schuster, K., & Jeschke, S. (2016). Educating engineers for industry 4.0: Virtual worlds and human-robot-teams: Empirical studies towards a new educational age. In 2016 IEEE Global Engineering Education Conference (EDUCON) (pp. 142–149). Abu Dhabi, UAE.

Samoila, C., Ursutiu, D., & Jinga, V. (2016). The remote experiment compatibility with Internet of Things. In 13th International Conference on Remote Engineering and Virtual Instrumentation (REV) (pp. 204–207). Madrid, Spain.

Sandu, F., Costache, C., & Balan, T. (2015). Semantic data aggregation in heterogeneous learning environments. In IEEE 21st International Symposium for Design and Technology in Electronic Packaging (SIITME) (pp. 409–412).

Sarıtaş, M. T. (2015). The emergent technological and theoretical paradigms in education: The interrelations of cloud computing (CC), connectivism and Internet of Things (IoT). Acta Polytechnica Hungarica, 12(6), 161–179.

Seeling, P., Roberts, S., & Weible, J. (2022). Towards education 5.0: Instruction with learners in the loop. In Proceedings of the 23rd Annual Conference on Information Technology Education (pp. 92–93). Chicago, Illinois.

Shi, Y., Qin, W., Suo, Y., & Xiao, X. (2010). Smart classroom: Bringing pervasive computing into distance learning. In Handbook of Ambient Intelligence and Smart Environments (pp. 881–910). Springer.

Shirehjini, A. A. N., Yassine, A., Shirmohammadi, S., Rasooli, R., & Arbabi, M. S. (2016). Cloud assisted IOT based social door to boost student-professor interaction. In International Conference on Human-Computer Interaction (pp. 426–432).

Srivastava, A., & Yammiyavar, P. (2016). Augmenting tutoring of students using Tangible Smart Learning Objects: An IOT based approach to assist student learning in laboratories. In International Conference on Internet of Things and Applications (IOTA) (pp. 424–426). Pune, India.

Sula, A., Spaho, E., Matsuo, K., Barolli, L., Miho, R., & Xhafa, F. (2013). An IoT-based system for supporting children with autism spectrum disorder. In 2013 Eighth International Conference on Broadband and Wireless Computing, Communication and Applications (pp. 282–289). Campiegne, France.

Sula, A., Spaho, E., Matsuo, K., Barolli, L., Miho, R., & Xhafa, F. (2014). A smart environment and heuristic diagnostic teaching principle-based system for supporting children with autism during learning. In 28th International Conference on Advanced Information Networking and Applications Workshops (pp. 31–36). Victoria, BC, Canada.

Tan, W., Chen, S., Li, J., Li, L., Wang, T., & Hu, X. (2014). A trust evaluation model for E-learning systems. Systems Research and Behavioral Science, 31(3), 353–365.

Tech, E. (2015). Internet of Things and the Humanization of Healthcare Technology. https://www.verizon.com/about/news/internet-things-and-humanization-healthcare-technology.

Tunc, C., Hariri, S., Montero, F. D. L. P., Fargo, F., & Satam, P. (2015). CLaaS: Cybersecurity lab as a service–design, analysis, and evaluation. In International Conference on Cloud and Autonomic Computing (pp. 224–227). Washington DC, United States.

Ueda, T., & Ikeda, Y. (2016). Stimulation methods for students' studies using wearables technology. In IEEE Region 10 Conference (TENCON) (pp. 1043–1047).

UNESCO. (2014). Teaching and Learning: Achieving Quality for All. http://uis.unesco.org/sites/default/files/documents/teaching-and-learning-achieving-quality-for-all-gmr-2013-2014-en.pdf

Uskov, V., Pandey, A., Bakken, J. P., & Margapuri, V. S. (2016). Smart engineering education: The ontology of Internet-of-Things applications. In IEEE Global Engineering Education Conference (EDUCON) (pp. 476–481). Abu Dhabi, UAE.

Uzelac, A., Gligoric, N., & Krco, S. (2015). A comprehensive study of parameters in physical environment that impact students' focus during lecture using Internet of Things. Computers in Human Behavior, 53, 427–434.

Valpreda, F., & Zonda, I. (2016). Grüt: A gardening sensor kit for children. Sensors, 16(2), 231.

Voas, J. (2016). Networks of "things." NIST Special Publication, (Vol. 800, p. 183).

Wan, R. (2016). Network interactive platform ideological and political education based on internet technology. In International Conference on Economy, Management and Education Technology. Chongqing, China.

Wang, J. (2015). The design of teaching management system in universities based on biometrics identification and the internet of things technology. In International Conference on Computer Science & Education (ICCSE) (pp. 979–982). Cambridge University, UK.

Wang, Y. (2010). English interactive teaching model which based upon Internet of Things. In International Conference on Computer Application and System Modeling (ICCASM 2010) (Vol. 13, pp. V13–587). Taiyuan, China.

Wang, Y. (2014). The construction of the psychological health education platform based on Internet of Things. In Applied Mechanics and Materials (Vol. 556, pp. 6711–6715).

Weber, R. H. (2010). Internet of Things–New security and privacy challenges. Computer Law & Security Review, 26(1), 23–30.

Xue, Y.-f., Sun, H.-l., Wu, Y.-h., & Chen, S.-k. (2011). Design and Development of digital teaching management system based on Internet of Things. In International Symposium on Computer Science and Society (pp. 138–141)..

Yin, C., Dong, Y., Tabata, Y., & Ogata, H. (2012). Recommendation of helpers based on personal connections in mobile learning. In IEEE Seventh International Conference on Wireless, Mobile and Ubiquitous Technology in Education (pp. 137–141). Takamatsu, Japan.

Zhang, N. (2016). A campus big-data platform architecture for data mining and business intelligence in education institutes. In 6th International Conference on Machinery, Materials, Environment, Biotechnology and Computer. Tianjin, China.

Zhu, L. (2016). Research and design of the future classroom based on big data and cloud processing. In International Conference on Audio, Language and Image Processing (ICALIP) (pp. 111–114). Shanghai, China.

11 Smart City in Action
A Case Study of the City of Santa Rosa, Brazil*

11.1 INTRODUCTION

Many cities around the world have developed and/or are planning to develop smart city programs. They hope to improve the quality of life of their citizens, increase the efficiency and effectiveness of urban operations, and develop solutions to overcome challenges, whether social, environmental, and from other natures.

The planning described in this chapter is applied to the city of Santa Rosa, a Brazilian municipality located in the northwest region of the state of Rio Grande do Sul, Southern Brazil. Figure 11.1 shows the geographical location of the Santa Rosa city in the country. The municipality was created as a colony of European immigrants, mainly Italians, Germans, and Russians in 1915. According to the Brazilian Institute of Geography and Statistics (IBGE), the estimated population in 2022 is 73,882 people, with a territorial area of 489,380 km², and a population density of 140.03 inhabitants per km². Its Municipal Human Development Index (IDHM) is considered high, with a value of 0.769 in 2022. The city's GDP per capita in 2022 reached BRL 43,564.15 (around US $8,712).

The city was traditionally known as "The National Cradle of Soy," as the first seedlings of the cereal were planted in Brazil exactly in that region. This title was made official by the Brazilian Congress in 2022.

The objective of this chapter is to present the infrastructure being developed by the authors at the Unijuí University and Federal University of Juiz de Fora, in collaboration with the public administration of Santa Rosa.

This project uses the LoRa communication network and provides global coverage for the urban and rural areas of the city. The public-use network will allow the interconnection of Internet of Things (IoT) devices and various sensors, storing the information collected in the local cloud.

11.2 THE STATUS OF SMART CITIES IN BRAZIL

"Ranking 2022 project" in Brazil is known as the Connected Smart Cities Ranking. It aims to determine the smartest cities in Brazil by evaluating them based on 75 indicators in 11 different thematic areas such as urbanism, economics, education,

---

* Gerson Battisti, Sandro Sawicki, Odaylson Eder, Pedro Henrique Dias Valle, Rafael Z. Frantz1, and Fabricia Roos-Frantz contributed to this chapter.

DOI: 10.1201/9781003348542-11

FIGURE 11.1 Santa Rosa geographical location in Brazil. (Adapted from Google Earth Pro.)

entrepreneurship, energy, governance, mobility, security, environment, technology and innovation, and health. The evaluation methodology uses values with three different weights (0.5, 0.8, and 1.0) to calculate the ranking and covers a range of factors, including smart traffic lights, smart lighting systems, electronic public transport tickets, risk area monitoring, appointment scheduling in the public health network, municipal legislation, computerized or georeferenced registration of properties, citizen services via smartphone applications, and others. The project was conceived by technology companies and provides a starting point to understand the indicators and levels of development of cities in Brazil.

According to the results of the 2022 Connected Smart Cities Ranking project in Brazil, the top ten cities are listed as follows: (1) Curitiba, (2) Florianópolis, (3) São Paulo, (4) São Caetano do Sul, (5) Campinas, (6) Brasília, (7) Vitória, (8) Niterói, (9) Salvador, and (10) Rio de Janeiro. The evaluation methodology took into account the five regions of Brazil (North, South, Midwest, Northeast, and Southeast) and found that the Southeast region dominated the ranking, with 58 out of the 100 smartest cities in Brazil located there. Additionally, six of the top ten cities were from the Southeast region, three from the South region, one from the Midwest region, and one from the Northeast region. The ranking also considered the population size of each city, with 11 cities having less than 100,000 inhabitants, 55 cities having a population between 100,000 and 500,000 inhabitants, and 34 cities having more than 500,000 inhabitants. Of the top ten cities, eight have a population of over 500,000.

An analysis of different criteria in Brazil reveals regional dominance in specific areas. The South and Southeast regions dominate in both health and mobility criteria, with the South having five cities and the Southeast having four cities in the top ten of both criteria. The technology and innovation criteria show a balanced distribution of cities among the South, Northeast, and Southeast regions. The economic axis is dominated by the South (three cities) and Southeast (seven cities) in the top ten positions. The population size of these cities varies, with some having less than 100,000 inhabitants, others between 100,000 and 500,000, and some having over 500,000 inhabitants. In general, the South and Southeast regions of Brazil have the largest concentration of potentially smart cities. Of the first 100 cities ranked in the general classification, 81 cities are in the South and Southeast regions, while the remaining 19 cities are in the North (2), Northeast (10), and Midwest (7) regions. These data also highlight social inequality, particularly in the North and Northeast regions of Brazil.

11.3 BUILDING LIVING LABORATORIES

The initiative to make Santa Rosa a smart city began with a public announcement by the Secretary of Innovation, Science and Technology of Rio Grande do Sul. The announcement selected three proposals for the creation of IoT and smart city experimentation laboratories, and the proposal led by Unijuí University in partnership with the municipal government and IT companies received funding of BRL 1,321,568.22. The resulting development laboratory, called Smart LiveLab, serves as a collaborative space for technological innovation to drive social and economic growth in the northwest and Missions macro region of the state. The Smart LiveLab consists of six environments that allow the four innovation pillars (society, business, government, and university) to utilize IoT technologies to create innovative, scalable, and sustainable solutions that have a direct impact on the macro-region's development. Table 11.1 presents the six environments of the Smart LiveLab project and their characteristics.

TABLE 11.1

Smart LiveLab Environments

Name	Description
Make LiveLab	In this environment, partners can carry out the prototyping process, pilot production, and component tests. The function of the space is to assist in the validation of the generated ideas. In addition, this space can be used for team training in recent technologies.
Idea LiveLab	Aimed at stimulating creativity and innovation. A pleasant environment associated with a set of technologies that include creative tools, computer equipment, and software to stimulate creativity and generate solutions that are in conformance with the context of the subject worked. The ideal place to emerge innovative ideas.
Storage LiveLab	Consists of a set of high-performance and high-capacity servers that allow the storage and retrieval of data collected by the sensors. Data is stored independently for each partner or test, ensuring the security and confidentiality of the information collected.
Front LiveLab	It addresses the reception of the partners and users of the laboratory, also constituting in a control and accounting of access of the users.
Collab LiveLab	Used for prospecting meetings for new partners and investments. Its composition is formed by the project coordinator, a representative from the private sector, a representative from universities, a representative from civil society, and another from the government. Its function is to create a network of partners around Smart LiveLab with a view to planning and prospecting the sustainable continuity of this project.
City LiveLab	The largest and broadest project activity includes the set of communication devices for IoT in an actual city environment. These devices, in addition to collecting and making public data from sensors available, allow partners to test and evaluate their products on communication issues, including distance, coverage area, security, QoS, among others.

Among the environments listed in Table 11.1, one of them stands out, as it is inseparable from the city itself. *City LiveLab* is intended to make the entire city a great laboratory for experimentation and testing in smart cities. This environment will be implemented in the city of Santa Rosa in partnership with the City Hall and the city's Secretariat for Economic and Social Development.

IoT and LoRa and two of the crucial concepts and technologies involved. IoT, as discussed in Chapter 2, refers to any device that can connect to the Internet through various connection types such as RFID, NFC, Wi-Fi, Bluetooth, ZigBee, or long-distance connections such as GSM, GPRS, 3G, and LoRa. LoRa, on the other hand, is a type of wireless network that uses radio-frequency waves to transfer small amounts of data over long distances. Communication can occur either through a Gateway or directly between devices. LoRa sets up a low-power wide area networks (LPWANs) created by Semtech to standardize LPWANs, LoRa provides for long-range communications: up to three miles (5 km) in urban areas, and up to 10 miles (15 km) or more in rural areas (line of sight). A key characteristic of the LoRa-based solutions is ultra-low power requirements, which allows for the

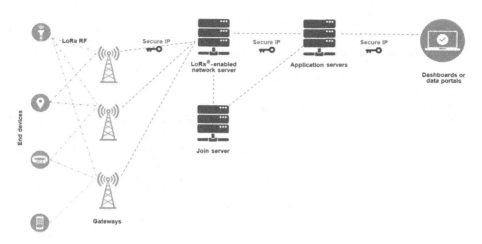

FIGURE 11.2   Typical LoRaWAN infrastructure and architecture.

creation of battery-operated devices that can last for up to 10 years. Deployed in a star topology, a network based on the open LoRaWAN protocol is perfect for applications that require long-range or deep in-building communication among a large number of devices that have low power requirements and that collect small amounts of data (SEMTECH, 2022).

Figure 11.2 presents a typical standard architecture for LoRaWAN. This figure demonstrates the integration between the main components involved in the configuration, which are end devices, gateways, network server, join server, application server, and dashboards. Table 11.2 details each of these components by exploring their functions (SEMTECH, 2022).

The main component, LoRaWAN, is the name given to the protocol that defines the system architecture as well as the communication parameters using LoRa technology. Thus, this protocol was developed by the LoRa Alliance, a non-profit organization founded in 2015. LoRa Alliance believes that the IoT time is now and that standardization and a strong, growing ecosystem is the only way to drive volume deployments for LPWAN. These LPWANs are designed to connect 50% of the predicted IoT volumes. The LoRa Alliance is standardizing LPWAN with the LoRaWAN specification and has created a certification and compliance program to ensure interoperability. LoRaWAN end-devices will be able to be deployed in multiple networks and roam from one network to another irrespective of network infrastructure or operator. The LoRa Alliance is the fastest growing technology. The alliance includes more than 500 members and has been operational since the end of March 2015. The members include technology leaders such as IBM, Cisco, HP, Foxconn, Semtech, and Sagemcom as well as the leading product companies such as Schneider, Bosch, Diehl, and Mueller and many SMEs and Startup companies all adding significant value to the fast growing LoRaWAN ecosystem. Our members also include the largest mobile network operators that are deploying public networks using the technology (LoRa Alliance, 2022).

TABLE 11.2

Typical LoRaWAN Architecture Component

LoRaWAN Architecture Component	Description
End devices	A LoRaWAN-enabled end device is the basic element or module of the network. They represent sensors or actuators that are connected by radio frequency signals to a network LoRaWAN through radio gateways using LoRa modulation. In the majority of applications, an end device is an autonomous entity and can be a sensor such as for measuring temperature, motion, energy consumption, water, gas, or actuators such as panic buttons, street lighting, and irrigation controllers, among others.
Gateways	A LoRaWAN gateway is the receiver of the signals sent by radio, a LoRa modulated RF message from any end device. A gateway can receive data from thousands of devices and forward them to a network server. In some models, the network server is integrated into the gateway. An external gateway, depending on the conditions of the site's topology and its antenna, can cover from 2 to 15 Km.
Network server	The LoRaWAN server guarantees authenticity of each sensor on the network and the integrity of each message. It comprises software running on a server, responsible for managing the information sent by gateways. Since there is the possibility of two or more gateways receiving the same packet from a certain module, the network server validates the authenticity and integrity of devices, deduplicates uplinks, selects the gateways used for downlinks, and performs adjustments to adapt the rates of data (data rate – DR) to manage the times between communications and the energy consumption.
Join server	The join server manages the activation process during transmission to add end devices to the network server. For that purpose, the join server must contain device information, such as DevEUI (End Device Serial Unique Identifier), AppKey (Encryption Key Application), NwkKey (Network Encryption Key), Application Server Identifier, and End-Device Service Profile.
Application server	Application servers are responsible for securely handling, managing, and interpreting sensor application data. They also generate all the application-layer downlink payloads to the connected end devices.
Dashboard or data portal	They are the means of displaying the data already processed and interpreted. Data can be viewed on mobile phones, web pages, and industry internal control panels, among others.

Source: Adapted Semtech (2022).

NETWORK PROTOCOLS

Network protocols are a set of rules that define how data are transmitted between devices over a network. They are necessary to ensure that devices can communicate with each other effectively and efficiently. In the context of IoT, network protocols are essential for connecting devices to the Internet and allowing them to exchange data with other devices and applications.

There are several network protocols used in IoT, including:

- **Message Queuing Telemetry Transport (MQTT):** A lightweight protocol designed for efficient communication between devices with low bandwidth or unreliable networks. It is widely used in IoT applications for machine-to-machine communication.
- **Constrained Application Protocol (CoAP):** Another lightweight protocol designed for use in resource-constrained environments. It is used in IoT applications to enable communication between devices and to transfer data over the Internet.
- **Hypertext Transfer Protocol (HTTP):** A standard protocol used for transmitting data over the Internet. It is widely used in IoT applications for communication between devices and web applications.
- **ZigBee:** A wireless protocol designed for low-power, low-data-rate devices. It is used in home automation and industrial control applications.
- **Bluetooth Low Energy (BLE):** A wireless protocol designed for low-power devices that require short-range communication. It is commonly used in wearable devices and smart home applications.
- **Long range wide area network (LoRaWAN):** A wireless protocol designed for long-range communication between devices with low-power consumption. It is commonly used in smart cities and agriculture applications.

The project to make the city of Santa Rosa a smart city considers the need to use the LoRaWAN architecture for reasons of compatibility and must be integrated into the spaces defined by the Smart LiveLab project.

11.4 MAKING THE CITY SMART

The smart city project in Santa Rosa aims to implement the LoRaWAN architecture, with the goal of achieving full compatibility and integration into the spaces defined by the City LiveLab initiative. To achieve citywide coverage, the project involves installing a sufficient number of LoRa gateways, chosen after careful analysis of available market options. The RAK brand equipment was selected for its compatibility with the other components in the network. The main components for LoRa

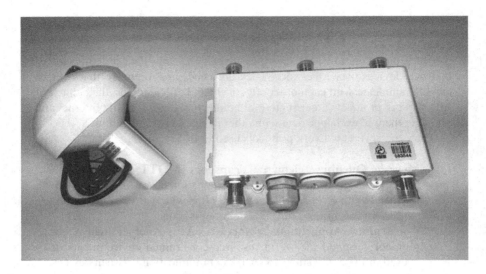

FIGURE 11.3   External gateway – RAK7240.

network coverage in the city include the external RAK7240 gateways (Figure 11.3), which feature eight channels with 4G, GPS Antenna, 2.4G Wi-Fi Antenna, and power via PoE injector.

The RAK7240 WisGate Edge Prime is an outdoor LoRaWAN gateway for LPWAN deployment. It is based on the SX1301 LoRa core. The main hardware and software characteristics are presented in the Table 11.3.

Signal range is a crucial aspect of the entire citywide coverage project using LoRa technology. To ensure optimal coverage, antennas with higher power and overlapping signals were selected. The 8dBi Fiberglass Antenna was chosen for use with the gateway, as depicted in Figure 11.4.

The 8dBi Fiberglass Antenna, designed for outdoor deployment, was selected for its improved signal range. The antenna body and connector are integrated to enhance its resistance to external conditions and its waterproof design (IP67 rated) makes it

---

TABLE 11.3
RAK7240 Hardware and Software Features

Hardware Features	Software Features
IP65 industrial-grade enclosure with cable glands	Built-in LoRa server
PoE + surge protection	OpenVPN
Dual LoRa concentrators for up to 16 channels	Software and UI sit on top of OpenWrt
Backhaul: Wi-Fi, LTE, and Ethernet	LoRaWAN 1.0.2
GPS	LoRa frame filtering (node whitelisting)
SD card slot	MQTT v3.1 Bridging with TLS encryption
	Buffering of LoRa frames in case of NS outage (no data loss)

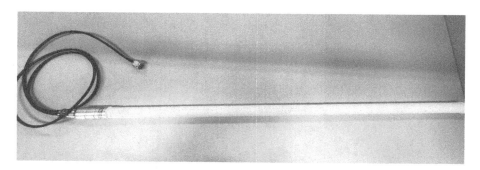

FIGURE 11.4   8dBi Fiberglass antenna.

suitable for outdoor use. The first configuration and installation step was carried out in the laboratory and a picture (Figure 11.5) shows the testing of the first external gateway.

The initial external gateway of the citywide LoRaWAN has been set up and is operational at the university. As shown in Figure 11.6, the antenna is placed on top of the building.

To store the values collected from the sensors, a Dell server will be used, more specifically the PowerEdge R550 model (Table 11.4). The VMs to receive data will be installed on this server. The basis of the system is Microsoft Hyper-V Server.

For the management of the various gateways and sensors, the ChirpStack software was defined. It is an open-source LoRaWAN server, join server, and application server, which can be used to build and manage private or public LoRaWANs. Figure 11.7 presents ChirpStack and a real-time view of the status of gateways

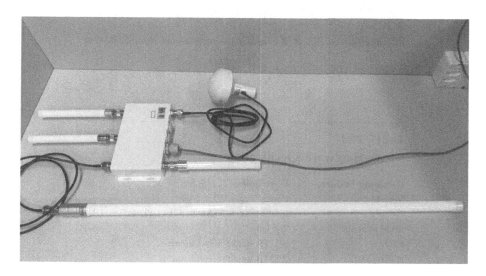

FIGURE 11.5   External gateway laboratory test.

FIGURE 11.6   First external gateway installed.

TABLE 11.4
Server Hardware Specification

Hardware Features	Description
Model	PowerEdge R550
Processor 01	Intel(R) Xeon(R) Silver 4314 CPU @ 2.40 GHz
RAM memory	256 Gb
SSD	1.77TB Raid 5 + Hot Spare
HDD	7.28TB Raid 5 + Hot Spare

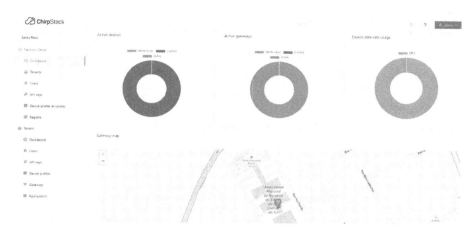

FIGURE 11.7   ChirpStack software for Santa Rosa LoRaWAN.

and sensors. The dashboard presented is the beginning of the configuration of the LoRaWAN covering the city.

11.5   COVERAGE AREA

A key aspect of transforming Santa Rosa into a smart city is ensuring that the necessary infrastructure is in place to reach all locations. To achieve this, the LoRaWAN must provide total coverage of the city's geographical area. The process of selecting locations took into account the involvement of the government entity (city hall), which made Basic Health Units (UBS) available for installing gateways. The UBS are basic healthcare facilities spread throughout the city, with a greater concentration in the central region but also present in rural areas, for a total of 18 units.

The previous section introduced the gateways and antennas that will be used in the project. Experiments have shown that they have a coverage range of up to 15 kilometers in radius. However, the range may vary based on factors such as topography, obstacles, trees, etc. To account for these signal interference factors, the distribution of gateways was planned to have a range of 7 km in rural areas and 3 kilometers in urban areas of the city.

The initial deployment of ten gateways is depicted in Figure 11.8, showcasing their estimated coverage areas. The larger circles represent installations with a coverage range of 7 km, while the smaller circles represent installations with a signal coverage of 3 km. The lines on the map represent the boundaries of the municipality. The potential overlap of signal coverage is advantageous, as it allows the sensors to transmit information to the gateway with the strongest signal, reducing energy consumption. The locations of the gateway installations and their respective coverage ranges are listed in Table 11.5.

This structure will allow the collection, transfer, and storage of information from sensors scattered throughout the city.

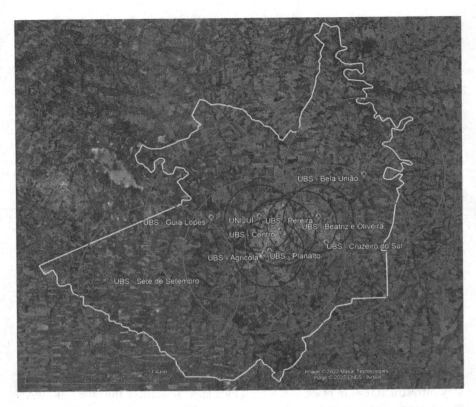

FIGURE 11.8 Gateway distribution map.

TABLE 11.5
Installation Point and Range

Installation Point	Range
UBS Agrícola	7 km
UBS Bela União	7 km
UBS Cruzeiro do Sul	7 km
UBS Guia Lopes	7 km
UBS Sete de Setembro	7 km
University Campus	3 km
UBS Beatriz e Oliveira	3 km
UBS Centro	3 km
UBS Pereira	3 km
UBS Planalto	3 km

11.6   SMART SERVICES UNDER DEVELOPMENT

The LoRa infrastructure proposed for the city of Santa Rosa consists of capturing data from different sensors compatible with the LoRaWAN protocol and transmitting the data through LoRa gateways that cover the entire city (urban and rural) to the server (storage), which is located at Unijuí University campus of Santa Rosa. This session presents four smart services that are in progress in partnership with the municipal government, undergraduate students in Computer Science and Software Engineering graduation courses, and master and doctoral students in the Postgraduate Program in Mathematical and Computational Modeling at Unijuí.

11.6.1   River Water Level Monitoring

The city of Santa Rosa experiences recurrent disruptions due to small rivers that overflow during rainy periods. This leads to inconvenience for residents living near the rivers, damage to merchants and public roads, and puts people's lives and physical well-being at risk. One such incident reported by the Civil Defense forced more than a hundred families to evacuate their homes due to flooding in the Pessegueiro and Pessegueirinho rivers. Neighborhoods like Vila Santa Inês, Sulina, Aliança, and Auxiliadora are particularly vulnerable to flooding during heavy rains.

To help mitigate this problem, the city will measure river levels in real time and transmit these data to relevant authorities and residents in flood-prone areas. The data will be stored to create a historical record of water levels throughout the year. This information will serve as a reference and trigger alerts to keep the public and the municipal government informed and prepared.

Khomp brand sensors will be employed to monitor the height of small rivers that flow through the city. The ideal monitoring points are shown in Figure 11.9. The

FIGURE 11.9   River water level monitoring map.

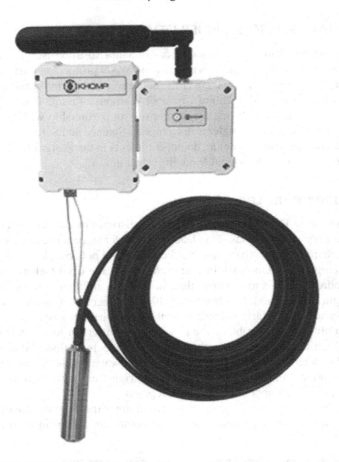

FIGURE 11.10   Level sensor for liquids.

liquid level sensor precisely measures the static pressure of the liquid in proportion to its depth by utilizing a high-quality piezoresistive diffuse silicon pressure sensor. Figure 11.10 displays the level sensor, which is suitable for liquids up to 20 m.

11.6.2   MONITORING OF AIR AND SOIL TEMPERATURE AND HUMIDITY

The 2021 Report on the Conjuncture of Water Resources in Brazil (CRHB) by the National Water Agency (ANA) states that demand for water in Brazil has been on the rise, particularly for the needs of cities, industry, and irrigation. The amount of water used for irrigation has increased from 640 to 965 m³/s over the past two decades, accounting for roughly 50% of total water usage in 2020. ANA predicts a 42% increase in water withdrawals over the next 20 years, rising from 1,947 m³/s to 2,770 m³/s, equivalent to an additional 26 trillion liters per year (ANA, 2022).

The report by ANA stresses the importance of planning for water usage to prevent future supply shortages. While it is true that proper watering of crops is crucial for their growth, it is also imperative to avoid wasting water resources.

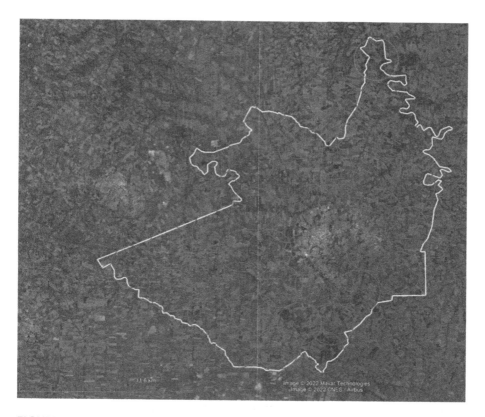

FIGURE 11.11   Map of air and soil temperature and humidity sensors.

A comprehensive monitoring system can support farmers in making informed decisions about when to irrigate their crops, as discussed in Chapter 9. This system constantly collects and consolidates data from soil humidity sensors and atmospheric sensors (temperature and relative humidity) on a daily basis.

To ensure uniform data throughout the municipality, the sensors will be placed evenly spaced, allowing for the creation of a soil temperature map covering the entire municipality. The plan calls for the installation of 40 sensors in rural areas, spaced approximately 4 km apart. Figure 11.11 shows the sensor distribution map.

The monitoring system will also be utilized in the urban area to manage the upkeep of the city's squares and parks. The collected information on soil temperature and humidity will aid in determining the necessary irrigation for trees and flowers. Figure 11.12 displays a representation of the soil temperature and humidity sensor.

11.6.3   Climate Monitoring

Monitoring the climate is crucial for various aspects of society, especially in agriculture, where it plays a crucial role in determining planting and fertilization schedules,

FIGURE 11.12   Soil humidity sensor module.

as well as in conserving water resources for irrigation purposes. The civil defense of the city can also benefit from the collected data in its planning efforts.

The city of Santa Rosa will be equipped with two meteorological stations to monitor the climate. As depicted in Figure 11.13, the stations will come equipped with sensors that gather data on wind direction, wind speed, precipitation, temperature, humidity, UV radiation, and atmospheric pressure.

To effectively monitor the climate in Santa Rosa, two meteorological stations will be set up. As shown in Figure 11.14, these stations will feature sensors capable of collecting data on wind direction, wind speed, precipitation, temperature, humidity, UV radiation, and atmospheric pressure.

The sensors in these stations are designed to operate optimally, and their location must be carefully selected to avoid any potential interference. As such, the proposed locations for the stations are the municipal airport and the university campus, allowing for the differentiation of climate monitoring in rural (airport) and urban (university) areas, as depicted in Figure 11.14.

11.6.4   Air Quality Monitoring

Air pollution is commonly linked to respiratory and cardiovascular illnesses. However, its full impact on human and animal health is yet to be fully understood. According to a study by Costa et al. (2020), pollution can have a negative impact on the brain and may contribute to central nervous system diseases. The authors

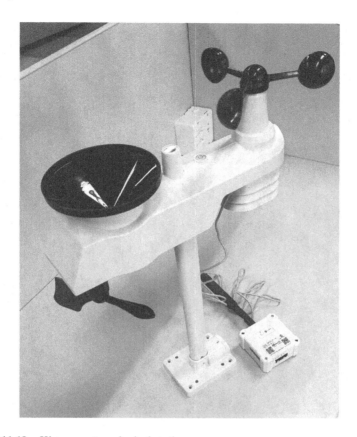

FIGURE 11.13   Khomp meteorological station.

mention various studies showing a correlation between exposure to air pollution and markers of neurodegenerative diseases such as Alzheimer's and Parkinson's, as well as a connection between air pollution and neurodevelopmental disorders including autism spectrum disorder.

Air pollution is composed of particulate matter (PM), which is a mixture of solid and liquid particles that are suspended in the environment. An important contributor to PM is traffic-related air pollution, mainly attributed to diesel exhaust. PM is classified into coarse, fine, and ultrafine, with PM2.5 being a method used to describe pollutant levels both in outdoor and indoor environments. PM2.5 are fine particles that have a diameter of less than 2.5 ms and remain suspended in the air for a longer time. Due to their size, they can travel deep into the respiratory tract, reaching the lungs and entering the bloodstream.

To measure air pollution levels in Santa Rosa, the city will use the Khomp brand NIT K72623-LO sensor, which is capable of detecting micro-particles (PM2.5). This sensor also monitors noise, temperature, and air humidity. Figure 11.15 displays the pollution monitoring sensor being calibrated in a laboratory before it is installed in

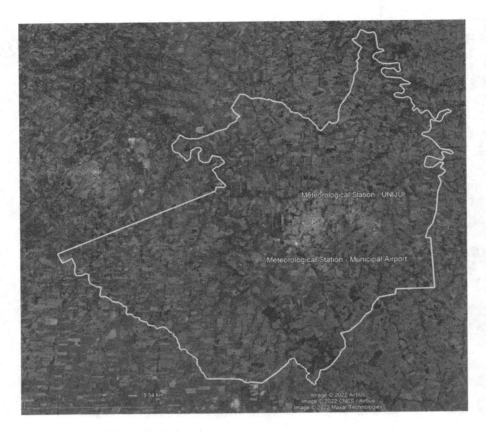

FIGURE 11.14   Meteorological station map.

FIGURE 11.15   PM2.5 sensor.

FIGURE 11.16   Air quality monitoring sensor map.

its final location. The sensor is powered by a battery and has a solar panel to keep it charged.

As the burning of fossil fuels by automobiles is one of the main causes of air pollution, the sensor will be installed in the central area of the city, next to the police station (Figure 11.16).

11.7   EXPECTED SMART SERVICES FOR THE NEAR FUTURE

The implementation of a LoRa infrastructure in the city of Santa Rosa provides the planning of new intelligent services to be developed for the near future. This section describes nine possible smart services that can be implemented in the municipality of Santa Rosa using the LoRa technology and sensors compatible with the LoRaWAN protocol.

11.7.1   SMART BUS TRACKING SYSTEM

LoRa technology can be utilized in several ways to monitor public bus transportation. For instance, in Nagoya, Japan, the location of buses was located using LoRa, GPS, and Arduino at bus stops (Boshita et al., 2018). LoRa networks with GPS and LoRa modules have been successfully implemented in cities like Buenos Aires and Córdoba in Argentina (Gríon et al., 2017). A public transport tracking system in India has also been proposed by James and Nair (2017) utilizing the LoRa network and GPS, with LoRa wireless transmission used for communication between bus stops and a base station.

Unexpected conditions, such as heavy traffic, weather, roadblocks, and mechanical issues, can affect the movement of buses. Knowing the exact location and estimated arrival time of buses based on normal traffic conditions can improve reliability in public transport. In Santa Rosa, public transportation is provided exclusively by bus, traveling through the city's neighborhoods on 60 predefined routes with a fleet of 43 buses. All routes and schedules can be found on the transport company's website.

The proposed infrastructure for Santa Rosa involves to install LoRa gateways throughout the entire city (urban and rural) to allow the buses to be equipped with integrated GPS and LoRa modules that capture their positions in real time. These positions will be transmitted through the LoRa gateways to a data server located on the Unijuí campus in Santa Rosa. This implementation of a bus tracking system allows for the real-time monitoring of bus locations from smartphones, computers, and mobile devices, ultimately optimizing wait times at bus stops and improving the organization of citizens' daily activities.

11.7.2 Smart School Bus Monitoring System

The infrastructure for tracking school buses and public transportation is similar. In addition to the already mentioned LoRa technology applications, there are several studies that specifically focus on school bus tracking. For instance, the study by Kanaan et al. (2018) proposes an integrated system for tracking and monitoring school buses at the Islamic University of Lebanon using LoRa technology. The aim is to optimize student wait time at bus stops and provide an estimate of the bus arrival time. Kadam et al. (2018) also proposes a system for tracking public buses using GPS and providing users with the estimated time of arrival through an Android app.

In Santa Rosa, there are 68 schools grouped into four categories: private, municipal, state, and federal. Of these, 11 are private, 23 are municipal, 32 are state, and 2 are federal schools. The municipal government only provides free public transportation to students at municipal schools through 11 buses on 14 routes. The buses pick up students from both urban and rural areas and drop them off at home after classes.

The implementation of a school bus tracking system in Santa Rosa can benefit both parents and students in rural areas. They can reduce wait time, especially on inclement weather days and parents can keep track of their children's return home, increasing family security. The proposed LoRa infrastructure for public bus transport monitoring can also be utilized for the school bus tracking system.

11.7.3 Intelligent IoT Systems for Traffic Management

Traffic congestion is a prevalent issue in urban cities with an increasing number of vehicles every year. Multiple sensors can help monitor the flow of vehicles on roads and estimate congestion levels. The world is testing and using various technologies and methods to detect vehicles at intersections. One such technology is LoRa, which can be used to monitor traffic congestion in a given area. LoRaWAN protocol compatible sensors can be accessed over long distances while consuming low power. Nor et al. (2017) suggested a congestion model that relies on readings from LoRaWAN protocol compatible sensors to detect the presence of vehicles and improve data accuracy. The collected data is transmitted through the LoRa network to a server.

In 2022, Santa Rosa had around 60,000 registered vehicles with the traffic depart-ment of the state of Rio Grande do Sul. Despite its relatively low population, the city has numerous intersections, traffic lights, and roundabouts, causing congestion during peak hours due to the lack of real-time monitoring of vehicle flow. Our pro-posal aims to monitor vehicle flow through LoRaWAN sensors, providing real-time information about congested areas. This will enable traffic light synchronization and optimized vehicle flow, reducing congestion. Drivers can also receive real-time updates on congested areas.

11.7.4 PUBLIC GARDEN AND PARK

The impact of climate change has led to the adoption of smart city concepts, which aim to conserve natural resources through the use of advanced technologies. As an example, in Frankfurt, Germany, a smart irrigation system for urban trees was developed using IoT and data analytics. With the help of specialists in municipal administration, IoT, and botany, LoRa technology was used to transmit sensor data from 18 sensors installed in eight trees (Gimpel et. al, 2021).

Santa Rosa, with its tropical climate and intense heat, often sees its parks' veg-etation suffer due to water scarcity. To address this issue, the city is proposing the implementation of real-time monitoring in its 44 parks. By installing soil moisture and temperature sensors, city managers can be alerted of the need for irrigation and automatically trigger a controlled and intelligent system based on soil conditions. This would allow for a more efficient use of water.

11.7.5 NOISE MONITORING SYSTEMS

Noise mapping is a valuable tool for visualizing and tracking noise pollution in real time in cities. Liu et al. (2020) highlight the benefits of implementing noise maps and the technical challenges associated with using mobile and acoustic sensors. The use of noise sensors can help municipal managers to distinguish various sounds, such as vehicle accidents, gunfire, distress calls, sound limits in cars, etc. Placing noise sensors in buildings, schools, and offices can enhance the security of these places. Furthermore, when combined with temperature and brightness sensors, noise sensors can help detect when lights or air conditioners are turned on, potentially reducing energy costs. At the Unijuí campus in Santa Rosa, a combination of noise, temperature, and light sensors that are compatible with the LoRaWAN protocol will be initially installed in classrooms to assist university managers in monitoring their environment.

11.7.6 SMART BIKE SHARING

As discussed in Chapter 7, the use of alternative modes of transportation, such as walking and cycling, can help reduce carbon emissions and improve quality of life in cities. Torres et al. (2021) proposed a bicycle tracking system using LoRa technology to encourage its use at the Polytechnic Institute of Viana do Castelo in Portugal.

Our proposal aligns with this approach but specifically targets the municipality of Santa Rosa, which has a large bicycle path network. The plan is to equip bicycles made available by the municipal government, private sector, or social organizations with low-cost GPS sensors. These sensors can transmit data to a server through the LoRa gateways infrastructure, allowing real-time monitoring of bicycle location, usage time, route, and speed.

11.7.7   SMART PARKING SYSTEM

The increase in vehicle ownership and traffic congestion is a significant challenge in modern society, especially in medium and large cities with a high concentration of commerce. Santa Rosa, for instance, has around 60,000 vehicles, leading to a scarcity of parking spaces during peak times.

A smart parking system can improve the situation by optimizing waiting time and making daily life easier for citizens. With a heat map displaying the number of available parking spaces, individuals can make informed decisions on whether to use their vehicles or opt for alternative modes of transportation. This not only reduces the emission of pollutants but also the risk of traffic accidents. The implementation of this system is feasible with the use of LoRaWAN ultrasonic sensors, as demonstrated by the works of Kodali et al. (2018) and Tung et al. (2019). The smart parking system in Santa Rosa can leverage the LoRa gateways' transmission infrastructure, covering the entire city.

11.7.8   SMART PUBLIC LIGHTING MONITORING

The implementation of smart meters and light sensors to monitor public lighting helps with easy measurement and control of electricity usage and lighting levels. This leads to more efficient spending by the public manager, as noted in studies by Sánchez et al (2020), Tung et al. (2019), and Muthanna et al. (2018).

Santa Rosa has 10,000 streetlights spread throughout the city. Currently, traditional streetlamps are switched on and off based on timers or day/night sensors. With such a large number of lamps, the maintenance team has a lot of work to do. However, it is often challenging to locate damaged lamps and notify the maintenance team, leading to dark streets and inconvenience for citizens. The goal of this project is to use real-time monitoring to identify damaged street lamps and reduce the waiting time for maintenance teams, thus improving the quality of service.

11.7.9   SMART PUBLIC WASTE CONTAINER MONITOR

Keeping modern cities clean is crucial. Overflowing waste containers in public areas create unsanitary conditions that can lead to various diseases. To address this problem, recycling bins and waste management can be upgraded using IoT technology. As seen in the work of Ziouzios and Dasygenis (2019), ultrasonic sensors and LoRa technology can be used to measure the waste level in each public bin.

Santa Rosa has 400 public bins spread across nine neighborhoods, classified between dry and organic waste collection. Knowing the location of each bin is also useful information for city management. Our project proposes the implementation of ultrasonic sensors with GPS, compatible with the LoRaWAN protocol, to monitor the waste levels in real time through the LoRa network. These notifications aim to improve waste collection efficiency and assist city management.

11.8 FINAL REMARKS

This chapter presented a smart city infrastructure being developed by Unijuí University in partnership with the public administration of Santa Rosa. This infrastructure is the result of the project that was granted BRL 1.321.568,22 to create a large development laboratory, called Smart LiveLab, through notice SICT n° 04/2021 TEC4B. The city of Santa Rosa was covered with LoRa technology provided for the installation of LoRa gateways. In short, the LoRa infrastructure consists in capturing data from different sensors compatible with the LoRaWAN protocol and transmitting the data through LoRa gateways that cover the entire city to the server (storage).

After the implementation of LoRa infrastructure in the city of Santa Rosa, new services can be planned and developed to comply with the demands of society. In this sense, this chapter proposed nine possible services to be developed, such as smart bus tracking system, smart school bus monitoring system, smart IoT systems for traffic management, monitoring system of temperatures for public gardens and parks, noise monitoring systems, smart bike sharing, smart parking system, smart public lighting monitoring, and smart public waste container monitor. Other services can also be thought, planned, engineered, and deployed using the LoRa infrastructure of the city.

Motivated by the technological advances of recent years, mainly in regard to sensors, the Internet, and artificial intelligence, it is possible to observe a tendency of cities to join the movement of becoming smart cities, receiving public and private investments to build such infrastructures. In that context, the city of Santa Rosa becomes another Brazilian city to have a LoRaWAN infrastructure, making it possible to provide smarter services to the population both in rural and urban areas. It is important to emphasize that this chapter described an inclusive project open to the whole society that has a clear roadmap of new services that will be added in the future to increase citizen participation, make the city more dynamic in terms of the use of resources, and improve aspects of health and safety of the city.

REFERENCES

Agência Nacional de Águas e Saneamento Básico (Brasil). (2022). Conjuntura dos recursos hídricos no Brasil 2021: relatório pleno/Agência Nacional de Águas e Saneamento Básico. ANA, Brasília.

Ashton, K. (2009). That "Internet of Things" Thing. RFiD Journal. Disponível em: http://www.itrco.jp/libraries/RFIDjournal-That%20Internet%20of%20Things%20Thing.pdf, Acesso em: 01/10/2021.

Boshita, T., Suzuki, H., & Matsumoto, Y. (2018). IoT-based bus location system using LoRaWAN. In 21st International Conference on Intelligent Transportation Systems (ITSC), Hawaii, USA.

Costa, L. G., Cole, T. B., Dao, K., Chang, Y. C., Coburn, J., & Garrick, J. M. (2020). Effects of air pollution on the nervous system and its possible role in neurodevelopmental and neurodegenerative disorders. Pharmacology & Therapeutics, 210, 107523.

Gimpel, H., Graf-Drasch, V., Hawlitschek, F., & Neumeier, K. (2021). Designing smart and sustainable irrigation: A case study, Journal of Cleaner Production, 315, 128048. https://doi.org/10.1016/j.jclepro.2021.128048

Gríon, F., Petracca, G., Lipuma, D., & Amigó, E. (2017). LoRa network coverage evaluation in urban and densely urban environment simulation and validation tests in autonomous city of Buenos Aires. In XVII Workshop on Information Processing and Control (RPIC), Mar del Plata, Argentina.

Harrison C. et al. (2010). Foundations for smarter cities. IBM Journal of Research and Development, 54(4), 1–16. https://doi.org/10.1147/JRD.2010.2048257

IBGE – Instituto Brasileiro de Geografia e Estatística. IBGE Cidades 2022. https://cidades.ibge.gov.br/brasil/rs/santa-rosa/panorama

Ismagilova, E., Hughes, L., Dwivedi, Y. K., et al. (2019). Smart cities: Advances in research—An information systems perspective. International Journal of Information Management, 47, 88–100.

James, J., & Nair, S. (2017). Efficient, real-time tracking of public transport, using LoRaWAN and RF transceivers. TENCON IEEE Region 10 Conference, Penang, Malaysia.

Kadam, A., J., Patil, V., Kaith, K, Patil, D., & Sham (2018). Developing a Smart Bus for Smart City using IOT Technology, 2018 Second International Conference on Electronics, Communication and Aerospace Technology (ICECA) (pp. 1138–1143). https://doi.org/10.1109/ICECA.2018.8474819

Kanaan, L. I. N. D. A., Haydar, J. A. M. A. L., Samaha, M. O. U. N. I. R., Mokdad, A., & Fahs, W. A. L. I. D. (2020). Intelligent bus application for smart city based on LoRa technology and RBF neural network. WSEAS Transactions on Systems and Control, 15, 725–732.

Kodali, R. K., Borra, K. Y., Sharan Sai, G. N., & Domma, H. J. (2018). An IoT based smart parking system using LoRa. In International Conference on Cyber-Enabled Distributed Computing and Knowledge Discovery (CyberC) (pp. 151–1513). https://doi.org/10.1109/CyberC.2018.00039.

Liu, Y. et al. (2020). Internet of things for noise mapping in smart cities: State of the art and future directions. IEEE Network, 34(4), 112–118. https://doi.org/10.1109/MNET.011.1900634

LoRa Alliance. What is the LoRa Alliance®? (2022). https://lora-alliance.org/about-lora-alliance/

Muthanna, M. S. A., Muthanna, M. M. A., Khakimov, A., & Muthanna, A. (2018). Development of intelligent street lighting services model based on LoRa technology. In IEEE Conference of Russian Young Researchers in Electrical and Electronic Engineering (EIConRus) (pp. 90–93). https://doi.org/10.1109/EIConRus.2018.8317037

Nor, R. F. A. M., Zaman, F. H. K., & Mubdi, S. (2017). Smart traffic light for congestion monitoring using LoRaWAN, In 2017 IEEE 8th Control and System Graduate Research Colloquium (ICSGRC) (pp. 132–137). https://doi.org/10.1109/ICSGRC.2017.8070582

Prefeitura Municipal de Santa Rosa. História. (2022). https://prefeitura.santarosa.rs.gov.br/?page_id=49

Ranking Connected Smart Cities. https://ranking.connectedsmartcities.com.br. 16/Oct/2022

Sánchez Sutil, F., & Cano-Ortega, A. (2020). Smart public lighting control and measurement system using LoRa network. Electronics, 9, 124. https://doi.org/10.3390/electronics9010124

Semtech, LoRa, and LoRaWAN: A Technical Overview. (2019). Disponível em: https://lora-developers.semtech.com/documentation/tech-papers-and-guides/lora-and-lorawan/, Access em: 01/09/2022

Sharifi, A., Allam, Z., Feizizadeh, B., & Ghamari, H. Three decades of research on smart cities: Mapping knowledge structure and trends. Sustainability, 2021, 13, 7140. https://doi.org/10.3390/su13137140

Torres, N., Martins, P., Pinto P., & Lopes, S. I. (2021). Smart & sustainable mobility on campus: A secure IoT tracking system for the BIRA bicycle. In 16th Iberian Conference on Information Systems and Technologies (CISTI). (pp. 1–7). https://doi.org/10.23919/CISTI52073.2021.9476495

Tung, N. T., Phuong, L. M., Huy, N. M., Hoai Phong, N., Dinh Huy, T. L., & Dinh Tuyen, N. (2019). Development and implementation of smart street lighting system based on LoRa technology. In International Symposium on Electrical and Electronics Engineering (ISEE). (pp. 328–333). https://doi.org/10.1109/ISEE2.2019.8921028

Ziouzios, D., & Dasygenis, M. (2019). A smart bin implementation using LoRa. In 4th South-East Europe Design Automation, Computer Engineering, Computer Networks and Social Media Conference (SEEDA-CECNSM). (pp. 1–4). https://doi.org/10.1109/SEEDA-CECNSM.2019.8908523

Index

Note: Locators in *italics* represent figures and **bold** indicate tables in the text.

Printed in the United States
by Baker & Taylor Publisher Services